Surveying sisters

Over the last few years there has been a dramatic increase in the number of women entering the surveying profession. Fewer than five per cent of practising surveyors are women, but women now comprise twenty per cent of students. *Surveying Sisters* explores the question of whether 'more' will mean 'better', either for women surveyors themselves, or for women as consumers of the built environment.

Clara Greed investigates the experiences of individual women surveyors, as well as studying the nature of the male majority. Taking a broadly feminist perspective and using an ethnographic approach, she develops a strong theoretical basis, incorporating the gender, class, and spatial dimensions of the situation, centring around the concept that surveying has its own distinct professional subculture. She traces the historical roots of the profession, and its attitudes to women, and makes constructive suggestions for improving the position of women in surveying today.

This is a highly topical study, at a time when the surveying profession is eager to attract more women in order to allay the effects of declining numbers of school leavers and potential 'manpower' shortages. It will be of interest to people concerned about issues of gender in disciplines such as sociology, management studies, higher education, urban geography, and women's studies, higher education, urban geography, and women's studies, and to the women and men who work in the surveying and the other built environment professions.

Surveying sisters
Women in a traditional male profession

Clara Greed

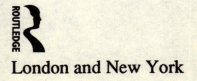
London and New York

First published 1991
by Routledge
11 New Fetter Lane, London EC4P 4EE

Simultaneously published in the USA and Canada
by Routledge
a division of Routledge, Chapman and Hall, Inc.
29 West 35th Street, New York, NY 10001

© 1991 Clara Greed

Typeset by Laserscript Limited, Mitcham, Surrey
Printed and bound in Great Britain by
Biddles Ltd, Guildford and King's Lynn

All rights reserved. No part of this book may be reprinted or reproduced or utilized in any form or by any electronic, mechanical, or other means, now known or hereafter invented, including photocopying and recording, or in any information storage or retrieval system, without permission in writing from the publishers.

British Library Cataloguing in Publication Data
Greed, Clara
  Surveying sisters.
  1. Great Britain. Women surveyors
  I. Title
  526.9082

Library of Congress Cataloging in Publication Data
Greed, Clara, 1948–
  Surveying sisters : women in a traditional male profession / Clara Greed.
  p. cm.
  Includes bibliographical references.
  1. Women surveyors   I. Title.
  TA534.G74   1990
  526'.0882—dc20

90–8210
CIP

ISBN 0-415-04845-1
       0-415-04489-8 (pbk)

Contents

List of figures and tables — vii
Preface — viii
Acknowledgements — ix

Part one   Surveying sisters? A study of the position and perceptions of women chartered surveyors

1 Is more better? — 3
2 Conceptual perspectives — 20

Part two   The historical perspective

3 The background to surveying up to 1900 — 43
4 Twentieth century development of surveying — 58
5 Women's presence 1945 onwards — 76

Part three   Education and practice today

6 The educational context — 89
7 Fitting into surveying education — 105
8 The position of women in surveying practice — 123
9 Getting by in the world of surveying — 143

Part four   Implications for the built environment and the profession

10 The influence of the subculture on what is built — 161
11 Conclusion — 180

Contents

Appendix one	RICS membership figures, 1989	200
Appendix two	Comparisons with other professions in 1989	202
Appendix three	A summary of the range of courses within surveying	205
Bibliography		208
Name index		227
Subject index		232

Figures and tables

Figure 1	The conceptual model	21
Figure 2	Realms of relevance	22
Table 1	Percentage of surveyors by division, 1986	123
Table 2	Percentage of JO membership by areas, 1986 (male and female)	127
Table 3	Percentage of all young surveyors in each sector	130
Table 4	Percentage of all women surveyors in each sector, 1988	130

Preface

The work of chartered surveyors covers all aspects of managing the built environment. Nearly 97 per cent of the profession is male. However, the number of women surveyors is increasing, and women now make up around 20 per cent of all surveying students in college. The study considers whether 'more' is 'better', either for the women surveyors themselves as fellow professionals, or for women in urban society as consumers of the built environment. The study is approached from a broadly feminist perspective using a qualitative 'sociological' style, but will be of interest to all those concerned with urban theory and professional practice. An ethnographic approach was adopted for the research, linking this to a strong conceptual basis centred around a study of the surveying subculture. First the historical development of the place of women in surveying is traced up to the present. This is followed by an account of the position of women in surveying education and practice today, illustrated by ethnographic insights. In conclusion, constructive suggestions are made for the future in order to use professional 'man'power more effectively, whilst enabling women surveyors to be both women *and* surveyors at the same time.

Acknowledgements

I would like to thank the following:

- Martin Joseph for allowing me to quote from his work on the professional socialisation of estate management students (1978, 1980) and for his interest and constructive opinions on my research.
- F.M.L. Thompson, whose work on the history of the surveying profession (1968) I frequently refer to, and for his interest.
- Mary E.H. Smith, past President of the Institute of Housing for her comments on the historical section of the book.
- The RICS for permission to use figures from the RICS Data Base on membership, as incorporated into Appendix I and Chapter 8, and especially Alan Cox, Assistant Secretary General, and J. Norris, Membership Director, for the useful information provided.
- All the surveyors, women and men, students and practitioners, who have participated in this research.
- Sandra Acker from the School of Education at Bristol University, and Linda McDowell, Senior Lecturer in Geography at the Open University, for their long suffering and guidance; Bristol Polytechnic, Department of Surveying, for giving me the time to undertake the study; and also John, Edie, Cathy, Joyce, Madge, Beverley, and Liz for their inspiration.

I acknowledge that Tables 1, 2, 3 and 4 incorporate material produced for the Salary Surveys (1986, 1988) by the JO (Junior Organisation of the RICS), and would like to thank the leadership and individual members for their assistance in this research.

The quotation from Macaulay's poem, *The Lays of Ancient Rome*, in Chapter 4 is reproduced courtesy of J.M. Dent.

Part one

Surveying sisters?
A study of the position and perceptions of women chartered surveyors

Chapter one

Is more better?

Introduction

Surveyors give professional advice on all aspects of land use and development, being particularly prominent within the private property sector. Their work is by no means limited to land surveying, but includes the valuation, investment, transfer, development, and management of land, and what is built upon it. Therefore the professional decisions surveyors make have a major influence on the nature of the built environment. This study is related to 'chartered surveyors', that is those who belong to the main professional surveying body, the Royal Institution of Chartered Surveyors (RICS).

This book is based on research I have carried out on the position of women in surveying, and the likely implications for women and the built environment which they inhabit (Greed, 1990a). Women comprise slightly over 3 per cent of the total fully qualified membership of the RICS, (rising to nearly 6 per cent if students are included, Appendix 1). The influx has been even more marked in surveying education where women now make up around 20 per cent of all estate management students. Taking the land use and construction professions and trades together, less than 5 per cent of those in practice are women (Greed, 1989). By contrast over 52 per cent of the population of Britain is female (Morphet, 1983) with over 80 per cent of women living in urban areas (OPCS, 1983). These factors may be irrelevant to urban policy-making, if women's needs are perceived as being no different from those of men; or if it is believed that the professional man is capable of sufficient disinterested neutrality to plan equally well for all groups in society (Dunleavy, 1980: 112). But as research and human experience have shown, women suffer considerable disadvantage within a built environment that is developed by men, primarily for other men (Stimpson *et al.*, 1981; Hayden, 1984; WGSG, 1984; Little *et al.*, 1988).

The increase in women entering erstwhile male professions

The last ten years have been marked by increasing numbers of women entering traditionally male-dominated professions such as accountancy, medicine, and law (Spencer and Podmore, 1987). The reasons for this growth are complex. On the one hand there has been increasing pressure to enter from women themselves, but the men have also welcomed this trend, albeit with caution, in order to compensate for the 'man'power crisis, caused by decreasing numbers of school-leavers (Robinson, 1987) and an expanding property market. Whilst the increase in the admission of women has benefited the professions, not least in providing 'new blood' and energy, many are concerned that this solution, unless coupled with fundamental organisational changes, is going to create an even greater crisis in the future (Law Society, 1988).

It is important to relate this study to material on what is happening in the other professions *vis-à-vis* women, and to keep in mind what one prominent woman town planner has called, 'the script for women in the professions in the Eighties' (Howatt, 1987). This study draws upon a wide range of material, including literature on women's relationship with science and technology (Swords-Isherwood, 1985; Whyte, 1986: Carter and Kirkup, 1989). Material from management studies was of interest to help understand how other professional and businesswomen got where they are in a man's world, and the problems they met on the way (Kanter, 1977; Marshall, 1984; La Rouche and Ryan, 1985).This cast light on the nature of the mechanisms that women encounter within the structures of the surveying profession, which may limit them in achieving their full potential. This study also draws on material concerned with the urban situation and the built environment, including work by urban sociologists, geographers, and of course surveyors and members of the other landed professions.

The urban question

One of the objectives of the study is to make a small contribution to answering one part of the classic urban question of 'who gets what, where, and why?' (Pinch, 1985; Diamond, 1986), and consequently to suggest recommendations for changing the situation. Its antecedents are in the literature of urban resource allocation, and more broadly urban sociology from a range of perspectives, albeit ungendered (Pahl, 1977a; Pickvance, 1977). It is certainly not the intention to reject this material as irredeemably biased because it is 'male' or at least 'malestream' (Siltanen and Stanworth, 1984: 186). Much of this work is drawn upon as a means of developing concepts for this study; but with caution. Indeed, the process is reciprocal, for example, one can now see echoes of feminist thought in the writings of the 'new' Pahl (1984). A problem that constantly confronts the feminist researcher in examining both urban literature and other relevant areas, such as work on the sociology of education (Acker, 1983a), is the question of whether the apparent silence on gender issues, in what is otherwise invaluable work, should be interpreted as meaning that women are included 'in' or 'out' of the discourse.

Spatial and aspatial factors

As the research developed, the emphasis shifted from looking at the land uses and developments themselves to looking at the aspatial (social) processes (Foley, 1964: 37) and value systems that determined the spatial end product. One of the conclusions of the research is that the spatial and aspatial cannot be separated, and that to make progress for the future, changes must be made in professional organisational structures to the advantage of women in order to facilitate the 'coming through' of alternative people and values so as to alter the nature of the policies, and the professions, that shape the built environment. Therefore both spatial and aspatial themes run parallel through this book. In particular, a discussion of the interrelationships between the surveying subculture, and gender, class, and space, forms an underlying sub-plot throughout. So, whilst considerable attention is given to the study of the profession and the position of women within it, this is couched within the context of a wider urban sociological discussion. Therefore it is often necessary to adopt a critical, theoretical, academic viewpoint towards the phenomenon of women in surveying, and to 'objectify' them (and the men); but this is balanced, at the personal level, by a measure of respect and admiration for the women and men surveyors I have met whilst undertaking this study.

Much of this study is concerned with identifying the subcultural values and processes that contribute towards 'the reproduction over space of social relations' (Massey, 1984: 16), especially the imprint of gender relations on space, that is, on the built environment. This process may be expressed in a simple thumb-nail sketch as follows, and will be elaborated in Chapter 2 and presented as a model, thus;

Gender, class, and other social factors → surveying subculture → space.

Both the horizontal and vertical distribution of women within surveying needed to be investigated as their relative position would determine 'who', male or female, would be in the right 'place' within the profession with the power to make policy decisions and thus shape the built environment. For example, not only were women under-represented in the higher levels of the profession but they were more numerous in certain specialist areas, such as residential work, than in others.

I became increasingly fascinated with the dynamics at work within the profession that determined which sort of people reach decision-making positions. Also, the identification and understanding of the values and assumptions which informed their professional opinions became a central pursuit in seeking to make sense of it all. For the purposes of this study it was helpful to see the world of surveying as a subculture. 'Subculture' is taken to mean the cultural traits, beliefs, and lifestyle peculiar to surveying. It is used both as a key theoretical concept, and a convenient shorthand term for describing the general milieu of surveying. One of the most important factors appeared to be the need for a person to be suitable, to 'fit in' to the subculture of surveying. It is argued that the values and attitudes of the subculture, as held by its members, have a major influence on their professional decision-making, and therefore ultimately influence the nature of urban development.

Gender is a major consideration in understanding 'who' receives what sort of treatment. Women as 'outsiders' have been particularly conscious of these mechanisms being used against them, but their experiences have often been dismissed as being too personal or 'emotional', and therefore of little real importance. But class as well as gender (to varying degrees in different contexts) proved to be a major fact in determining 'who' was considered suitable and 'the right type' within the surveying subculture.

The need for identification with the values of the subculture would seem to block out the entrance of both people and alternative ideas that are seen as 'different'. The concept of 'closure' as discussed by Parkin (1979: 89–90), and first developed by Weber (1964: 141–52, 236) in relation to the power of various sub-groups protecting their status is a key theme. This is worked out on a day-to-day basis at the interpersonal level, with some people being made to feel awkward, unwelcome, and 'wrong'; and others being welcomed into the subculture, made to feel comfortable (Gale, 1989a and b), and encouraged to progress to the decision-making levels within it. It is a major hypothesis of this study that one should not see all the 'little' occurrences of everyday (i.e. the encouragements and discouragements, nicenesses and nastinesses) as being trivial, irrelevant, or not serious enough to be counted as real data for the research; but rather as the very building blocks of the whole subcultural structure.

The development fraternity

Of course there are other actors, apart from surveyors, involved in the development process (Kirk, 1980: 38–41; Ambrose, 1986: 68–9), including architects, town planners, and councillors; and also the financial institutions and developers themselves. Significantly, most of these groups are predominantly male. Surveyors cannot be 'blamed' for everything, as their power varies according to the particular situation, and, relatively speaking, they differ among themselves and not all are directly involved in planning and development. It is not possible to absolutely isolate either surveyors, or women in surveying, from the surrounding societal and professional context in which they operate. Nor can one absolutely 'prove' the precise influence of the surveying subculture on urban society in general, or its effects on a specific development in particular. However, relatively speaking, the situation in surveying is typical of that pertaining in other landed professions, and so many of the observations made in this study might be applicable to them too. Also, it was valuable to make comparisons between surveyors and their near neighbours, the town planners, within the territory of the landed professions, to bring out the specific characteristics of those that belonged to the surveying tribe.

Antecedents to this study

There have been very few studies of surveyors themselves, as against studies of urban processes, or other landed professions, with the notable exceptions of Michael Thompson's comprehensive historical study of the profession (1968), and Martin Joseph's valuable work on the professional socialisation of estate management students (1978, 1980, 1988: Chapter 16). I fully acknowledge the value of both of these excellent sources in undertaking my research. Women are scarcely mentioned in these works (but in fairness there were very few women in surveying at the time of these studies). Surveyors have occasionally produced reports about themselves, but except for work by the JO (Junior Organisation for surveyors under the age of 33) (JO, RICS, 1986, 1988) specific attention is not usually given to women in the profession. Also a series of conference papers on women land surveyors in other countries, including Bulgaria, Nigeria, and Finland, was produced for the International Federation of Surveyors conference in 1983 (FIG, 1983). More broadly, it is important to ask why both feminists and socialists have put considerable emphasis on studying the other urban decision-makers, and 'managers' (Pahl, 1977b; Bassett and Short, 1980) in the public sector of town planning and housing, whilst relatively neglecting the power of those landed professions working for the private sector (with some notable exceptions, Marriot, 1989; Ambrose and Colenutt, 1979; Simmie, 1981) from whence the initiative for much development derives in the first place.

Twenty years ago, Anselm Strauss (1968: 316) commented that no sociologist had written about, what he called, the odd and unusual professions of realtors and investment brokers in the USA. It would seem that this has not yet been fully rectified on either side of the Atlantic. This is in marked contrast to other professions such as law, management, and medicine. In particular there is a large North American literature on professional socialisation of men in medicine, the more ethnographic examples in the genre of Becker *et al.* (1961) being of most interest. In recent years, this has been joined by a small but growing area of work by women on women medical students, such as Lorber (1984), which has demonstrated that findings related to men cannot be indiscriminately applied to women in the 'same' situation.

The findings of such studies whether feminist or 'malestream' are not necessarily transferable across the Atlantic. Surveying is an especially 'English' professional monopoly which developed specifically to serve the needs of the landed interests in society. Land as a representation of power is particularly strong in English society, possibly because of its obvious scarcity within an island setting. Surveying does not manifest itself in quite the same form in other countries, the professional cake being divided up variously between civil engineers, architects, town planners, and others. The historical and contemporary scope and nature of the profession are major determinants of women's potential role as surveyors. The profession is still undergoing change: for example, it is currently having to examine its *raison d'être* in view of the 'threat' of 'harmonisation' with the rest of Europe in 1992 (DTI, 1988).

The surveying spectrum

The RICS is the largest and oldest of the professional bodies concerned with land use and development in Britain (Thompson, 1968), significantly many times larger than both the other main landed professions, namely the Royal Town Planning Institute (RTPI) and the Royal Institute of British Architects (RIBA) and arguably commensurately greater in its influence on the nature of the built environment (Appendix 2). It is not one profession but many, comprising within its scope many different specialist sections (called 'divisions'), practising within a diversity of professional fields (RICS, 1987a). Although a surveyor is often seen by the general public as a man standing at the side of the motorway, wearing an orange anorak, and holding a surveying pole or theodolite, this is a totally false, but significantly male image (compare with Menzies, 1985). Such men are more likely to be surveying technicians.

The surveying spectrum (RICS, 1986a; Greed, 1989) extends from technological areas where few women are found (for example there are only six women in minerals surveying in the whole of Britain) through to the quasi-technological areas such as quantity surveying in which about 3 per cent of practitioners are women, and across to more upmarket, commercial areas such as estate management where about 20 per cent of students and nearly 10 per cent of practitioners are women (JO, RICS, 1986: Appendix 1). A high proportion of women are also found in residential estate agency work within the private sector, not all of whom are professionally qualified, this being seen as a low status area by many women and men in the profession. At the other end of the spectrum, the smaller, more socially-oriented areas of practice are found, such as housing management where up to 50 per cent of students are likely to be women. There is considerable debate amongst surveyors themselves, as to whether housing managers count as 'real' surveyors. However, many women practitioners believe that housing is currently being recolonised by men who apparently perceive it as an appropriate area for male power and control that has to be freed (as they see it) from its reformist roots and the historical influence of 'lady bountifuls'. Such are the contradictions and complexities inherent in the world of surveying.

The role of the surveying subculture as producer, reproducer, or transmitter of values, is open to much debate. The questions of what its actual boundaries are, and whether it contains within it significant subdivisions were to prove of great importance in understanding the particular position allocated to women within the world of surveying. Of course, the surveying subculture cannot be looked at in isolation but must be considered in relation to the wider social, economic, political, and spatial context of contemporary Britain.

Throughout its history, the boundaries of the profession have fluctuated to include or exclude certain groupings for a variety of reasons. Surveyors are great pragmatists and survivors, flourishing under both Labour (Eve, 1948) and Conservative governments with equal ebullience. However, surveying is generally accepted to be a 'conservative' profession in all senses of the word, and yet at present it would appear to be in the interests of the profession to admit women. The reasons for this apparently enlightened attitude required further investigation. Of particular relevance is the question of what sort of women are attracted to, and accepted into surveying. Nowadays it would seem that women who subscribe to the values of the successful businesswoman (Hertz, 1986), or bourgeois feminist[1] as she is sometimes called, are most likely to be 'the right type'. This was to became one of the key observations that developed in the course of this study regarding understanding the women themselves. The rise of the bourgeois feminist (not a derogatory term) is linked to wider political changes in Britain in the 1980s (Greed, 1988). Not only have the numbers of women increased but the whole profession has expanded numerically, changed in orientation, and grown in prestige and status under the enterprise culture created by the Conservative government.

The nature of the woman surveyor: a foretaste

It should not be assumed that the women entering the profession will necessarily hold different views from the men; nor should any automatic assumption be made that they are likely to be feminist (ironically some men surveyors assume they are and feel threatened by them without good cause), or that they will become radicalised by their experiences of being in a minority in a male subcultural group. It is a fascinating question as to why some women surveyors have become feminists, and others have not, when they appear to have had similar life experiences. Regardless of their individual views, many women and men surveyors would argue that a surveyor's personal beliefs are purely a private matter, that should in no way influence 'his' professional judgement. Even if women's views are 'different' upon entering the profession, such are the powers of professional socialisation within the landed professions (Gibbs, 1987) that they are likely to change. Many do nevertheless retain an alternative viewpoint, and a more liberal perspective. Some even possess a highly developed 'double-consciousness' (Rowbotham, 1973) and thus are able to objectify and discuss their experiences with great insight.

Relatively speaking, women in the more private-sector-oriented aspects of the landed professions, especially surveying and to some extent architecture, are likely to be fairly 'conservative', and may appear, superficially, to have no conflict with the world of men and business. Indeed non-surveying women have commented to me that some women surveyors appear to them incredibly 'straight' and uncritical in their world view; and I have observed that those that exhibit these characteristics the most strongly are often the most successful, declaring, 'I don't think about it, I just get on with my job'. Such women are more likely to be motivated to enter the male professions by the prospect of future achievement, than in righting past wrong or changing the built environment. Some may simply see it as quite usual for women to go into business, and may have no argument with the world of men, 'I simply don't know what all the fuss is about'. Curiously they might assume that their own views reflect quite normal Conservative policy, as naïvely they seek to apply principles of individual achievement and business success (which really may be intended for men) to their own lives, which is in practice a very radical act.

On the other hand, women planners, and some women geographers, certain groups of women architects (such as Matrix, 1984), and many housing managers are more likely to be critical or even radical in their world view. They may campaign for the provision of better child care, especially creches, and greater state intervention on behalf of women in society, which may actually embarrass some women surveyors, 'they're always going on about creches, and showing us all up'. However, there is a gradation within the more radical groups between those who are strongly 'alternative' and even separatist in their politics, to those colloquially described as 'femocrats' (Leoff, 1987: 14) who seek to use existing governmental institutions to carry out change. Such women often express great faith in a non-confrontational approach based on 'reason' and negotiation with men in order to create alternative management structures to the advantage of women. The latter group includes a few women surveyors. Clearly women in the landed professions are not a unitary group any more than are the men. Such differences of view are of great importance for the women themselves in working out who are 'people like us'. It would seem that the differences perceived by men between each other, which contribute to the mechanisms of social closure within professional groups, are mirrored to some extent amongst the respective female groups. I was attracted to Berger and Luckman's idea of subuniverses (1972: 102) to explain these fine nuances within the subcultures of the landed professions.

Many women surveyors seem quite alienated from feminism and may never have read any feminist literature (or have any knowledge of the realms of academic sociology, or the humanities in general). Yet they possess some measure of feminist consciousness of their own, but are unlikely to express themselves in feminist jargon. They may not identify 'patriarchy' as the cause of their problems, or even think in terms of macro-sociological first causes (but undoubtedly experience their effects). They are more likely to see their problems as being personal, and either their own fault or that of those who work with them. They may be put off by the false media image of feminism and 'the way feminists dress and carry on'. Since women surveyors are operating in a minority situation, many consider it unwise to draw attention to themselves, or to openly discuss their problems and negative experiences, so as not to antagonise the men, possibly in the hope of achieving more in the long run. It's a matter of 'heads under parapets' (SBP, 1987: 4). When I talk to such women about their experiences they may start by saying, 'I'm not a feminist, but . . .'.

Approach to the study

In order to investigate the position of women, it was necessary to study both the men and the women in surveying. Obviously in a profession which is 95 per cent male, the men no doubt have the greater influence, through their professional activities, on the nature of the built environment. An investigation of how (or indeed if) men in the landed professions perceive women, and their needs as members of urban society and thus potential recipients of urban goods and services, was therefore essential. Their perceptions influence the extent to which women's needs are taken (seriously) into account in the development of urban policy; therefore attention was given to understanding the male backcloth against which the women play out their professional roles.

A considerable proportion of the empirical reportage included in this book is related to conversations with the women themselves, for the following reasons. First, at a time when significantly greater numbers of women are entering the profession, it was vital to investigate the question of whether 'more' women would mean 'better' or 'different' in respect of the possible influence they might have on the nature of the profession and thus on the policies deriving from it (Greed, 1988). It was important to investigate their motives for entering the profession, and to understand 'what made them tick'. Second, it was found that although there was a number of older women already in the profession, their influence did not appear to be commensurate to their numerical presence. It was necessary to look more carefully at the mechanisms at work within the profession which might be limiting women's role and progress. Talking to the women themselves produced alternative insights from 'below' as to how they experienced the surveying profession, and what was 'really' happening within it.

Emphasis has been put on listening to individual women at the personal level, as well as hearing what the all-pervasive male voice, present throughout the subculture, was saying. Many women naturally presented their view of surveying from the perspective of their own personal experience (Greed, 1990b). As the study developed, such personal accounts became a key empirical component of the research. Although I talked to a number of men surveyors too, a special comparative study of them was not made. The whole ethos of the professional world in which the researcher was operating was predominantly male and already 'known' so one could not help but make a subconscious comparison with the male reality as 'the norm' against which the women's experiences were measured. Likewise Gallese in her study of female graduates of Harvard Business School (1987:20) states that she found it virtually impossible **not** to compare the progress of the businesswomen with that of their male colleagues.

One cannot 'prove' absolutely that 'gender' does determine the nature of development, but one can at least illustrate the situation through material from empirical observations. However, nowadays it would appear to be more acceptable to aim at the 'empirical illumination of theoretical concepts' as Philip Cooke (1987) comments (with some reservations) from within the realms of urban geography. Indeed such is the nature of macro-sociological theory that is concerned with causation that it may be unprovable, as Sandra Acker (1984: 36) points out with respect to the sociology of education which is also one of the key areas of relevance to this study.

The question was how to go about observing the values and attitudes that shape the world of surveying. Out in the 'real world' of practice the values of the subculture were rarely overtly stated as everyone took them for granted, and rarely articulated them or wrote them down. However, in the educational situation they had to be more openly expressed in order to be transmitted to the next generation, and therefore they could be more easily observed. A predominantly qualitative 'sociological' approach (Mills, 1978) with substantial use of ethnographic methods (Hammersley and Atkinson, 1983) was adopted. This tied in with the overall style of the research and its antecedents in the study of professional subcultures, and its tenuous links with what may be broadly described as 'the Chicago school of sociology'. The study of the urban situation and the investigation of subcultures within it, using ethnographic and anthropological methods, and the development of a variety of theories of human interaction were all elements from this source which were of value to the research (Bulmer, 1984).

The position of the author in the study

In the tradition of feminist research it is more acceptable, even obligatory, 'to leave the researcher in' (Stanley and Wise, 1983). Therefore I will now declare my position and interest in the research. I am a town planner and member of the RTPI, who has in the past worked in local government, and who is a lecturer in a Polytechnic Department of Surveying in the provinces in the South of England. I am also entitled to call myself a town planning surveyor by dint of membership of the ASI (Architects and Surveyors Institute, which is significantly a non-RICS professional surveying body)[2]. So in many respects I am a person in surveying myself, although I also retain at least two other sets of double-consciousness, both as a woman and as a town planner (Du Bois, 1983: 111), giving me a broader perspective as a relative 'outsider'.

I became interested in the research topic for three main reasons. As a lecturer, I was fascinated by the question of whether the observable increases in the number of women students would have an effect on the nature of surveying education and the profession itself. Second, as a town planner who had become a feminist of sorts several years after qualifying, I was keen to apply my new found consciousness to my own field. Third, I simply sought to understand my own life, particularly in relation to my experiences of education, the built environment, and the landed professions. I was curious to compare my experiences with those of others in similar circumstances.

So I am part of what is being studied (Lury, 1987) and cannot be separated from it, as 'in this process, I too am subject' (Mulford, 1986). I am in a sense both researcher and researched. It therefore seemed artificial to pretend that the researcher was separate or above 'his' subjects as in much male research which is written in the passive voice even when he was a major actor in it (for example, Blowers and Pepper, 1987). I do not pretend to have an objective neutral approach and indeed would question the value of having one (Morgan, 1981). Gardner states (1976) that 'participatory research may lead to subjective richness but reduce objective accuracy', but I am purposely aiming at subjective accuracy.

Methodological approach

False images of women surveyors abound, and surface reality is not to be trusted. For example, a large property company used a photograph in one of its advertisements of a woman surveyor holding a map standing with two men on a building site, implying a third of the firm's professional on-site staff are women. I phoned the personnel officer to find out who this woman was: 'Oh they're all actors, real people don't look right, they look awkward and don't stand properly ... no I don't think we've got any real women that do that sort of work, I'll check on my computer.' I was made to feel silly for being so foolish to imagine she might be a real surveyor!

I soon learnt not to go by facts and/or appearances without further investigation. I found an ethnographic approach to the research the most useful in getting at 'the truth'. There are a range of interpretations as to exactly what ethnography is, and how it should be done (Woods, 1987). For the purposes of this research, it is interpreted as a method based on going into the subgroup or tribe of surveying and observing everything, in order to make sense of it all (Hammersley and Atkinson, 1983). Three ethnographic approaches were adopted. First, a selective ethnographic study of my own department of surveying was undertaken over a period of approximately three years. My observations were chiefly drawn from the perspective of teaching town planning and courses on the social aspects of planning and development, although I was also aware of the nature of the other areas of surveying education. Second, dispersed ethnography, which involved contacting a range of surveying practices and visiting other colleges, was carried out. This is a method which is not unusual in studies of the professions where people are distributed in a series of dispersed practices (Smart, 1984: Chapter 7). Altogether I contacted around 250 women surveyors, speaking to them either individually or in groups. Third, since I am, in a sense, a 'woman in surveying' myself, retrospective ethnography is used to develop sensitising concepts from my own life experience (Okely, 1978; Purvis, 1987: Hammersley and Atkinson, 1983: 179). I am triangulating between the three sources. Triangulation, incidentally, is both a surveying (Hart and Hart, 1973: 324) and an ethnographic term (Hammersley and Atkinson, 1983: 198).

Whilst trying to retain the basic ethnographic principle of remaining open to everything, my ethnography was directed towards four substantive issues: (a) the overall world view of surveyors as to what they see as obvious and right, (b) their assumptions about land use and development, (c) their assumptions about different types of women and 'people', and (d) what they see as the right type to be a surveyor. There is another subdivision running right through the research – between observing aspatial (social) processes within the subculture and observing surveyors' views on spatial land use and development issues.

When talking to women surveyors, I only had to ask them, 'why did you go into surveying?' and they were under starter's orders and off, with no stopping them. These were the points I wanted to find out about, which I would turn into actual questions if (on the rare occasion) we didn't cover everything in the course of normal conversation.

who they were
why they went into surveying
where they studied
what they were doing now 'generally' and 'exactly'
what it was like (no prompting good or bad)
whether they considered that their/women's attitudes to professional practice and to land use and development policy were different from men's
how they saw themselves in five/ten years time (with no prompting as to whether I was referring to their personal or professional life)
anything else they thought was important (not necessarily fishing for 'problems')

Style

Whilst a predominantly ethnographic approach to the research was used, it is not the intention to present the findings as a comprehensive traditional ethnographic report. Rather, ethnographic observations and insights are incorporated into the study in order to develop and highlight themes and linkages from the conceptual basis, and to illustrate the position of women in surveying – which is relatively uncharted territory from a sociological perspective. The choice of illustrations is based on a distillation of the most representative examples from my three research sources outlined above.

Therefore this book includes a considerable amount of soft data, comprising observations, anecdotes, examples, and reportage of what women (and men) surveyors have said, felt, and experienced, backed up with material from journals and other literature from surveying education and practice. Verbatim comments from my observations will be inserted throughout the text as unattributed quotes in single quotation marks. (Quotations from written sources will also appear in single quotation marks followed by acknowledgement of source.) Single words in quotation marks 'thus' are either metaconcepts or words used in a particular sense, such as in the title of the book '*Sex*' at '*Work*' (Hearn and Parkin, 1987, to which reference will be made briefly).

I have included accounts of the 'bad', as well as the 'good', aspects of the experiences of women and attitudes of men within surveying. Whilst some women may see this as counterproductive at a time when apparent 'progress' is being made, others, including myself, believe this is necessary, in order to deepen understanding of what is really happening within the profession from both an academic and practical perspective, to show newcomers that it is not 'just them'; and also to develop informed strategies for the future. Indeed, it must be pointed out that most of the observations recorded in this book have happened within the last few years, and some are still ongoing. Because of the potentially 'sensitive' content it seemed wise not to identify the specific source of any observation. However, it would be totally inaccurate to assume that, for example, all illustrations from education relate to the author's own college, or that those from professional practice relate to a particular locality, as in fact a wide range of sources was drawn upon through the various ethnographic routes identified above.

I decided, consciously, to use a more personal colloquial style in certain places, verging on the stream of consciousness genre found in certain areas of feminist literature, if it were thought to contribute to the impact of the narrative either in terms of illustration of observations or enhancement of analysis and reflection. Whether this is a sign of specifically 'feminist research' or 'feminist academic literature' or an attempt to communicate the research findings more effectively and directly is open to debate. I realise this approach would have been unthinkable only a few years ago, but is now becoming more acceptable, at least in some feminist academic circles.

Is more better?

I was often accused of being 'over-essentialist' (i.e. 'all lads are bad') in assuming that the 'problem' was 'the men'. In fact, I was well aware of the complexities of the whole situation, but would still hold that patriarchy must take much of the blame. However, I had no intention of indulging in a tautological analysis (as some imagined) in which one assumes the surveying subculture is essentially 'masculinist' so that any observations of the activities or values of the male members of that subcultural group are taken as empirical evidence of that male bias and as necessarily 'bad'. As one man put it, debatably, 'you can't blame them for being men, that's the way they are'. Indeed, some men appear more open to gender issues and will readily admit that there is something especially 'male' about the landed professions, over and above the situation in other professions or in society as a whole. For example, Dorfman (1986) comments that, 'Overly concerned with protecting and developing professional status, the RIBA, RTPI and RICS have become masonic strongholds forbidden to ordinary men, and women and ethnic minorities in particular'.

In this book, I will argue that gender does matter, but will also acknowledge the complexity of the interaction of gender with class, and the importance of other factors within the surveying subculture and within individual surveyors' lives. Whilst, in places, the book is actually complimentary towards surveyors, it should be seen as a critical sociological enquiry from a broader, urban feminist perspective.

Contents

Chapter 2 gives an outline of the conceptual basis that informed this study, material being drawn from a wide range of relevant academic realms (skip this chapter if you aren't academically inclined). Three chapters (3, 4, 5) are devoted to the historical development of the surveying profession with particular reference to the position of women within it. The first chapter (3) deals with the period up to 1900. Emphasis is put upon understanding the origins of the profession which contributed towards the shaping of its present-day value system. The roots of its noticeable veneration of 'Land' (Thompson, 1968; Joseph, 1980), and the position of women in dynastic surveying families, are examined. The following chapter (4) examines the development of surveying in the twentieth century up until the period following the Second World War. Women were first allowed to enter the male professions following the 1919 Sex Disqualification (Removal) Act. This led to a small cloudburst of women surveyors, but this was followed by a subsequent drought which persisted right up until the relative flood of female entrants in recent years. The remaining historical chapter (5) looks at the development and growth of the modern surveying profession, and the manifestation of a second wave of women entering the profession from the 1960s onwards, a trend which has much accelerated recently.

The next section deals with the position of women in surveying education and practice today, illustrated by ethnographic observations. This is subdivided as follows: Chapter 6 gives an account of the current structure of surveying education and women's position within it. The following chapter (7) gives a more sensitive account of women's experience of surveying education, both chapters drawing extensively on ethnographic material. Particular attention is given to surveying students' attitudes towards women and their needs, and to material regarding their assumptions about the ideal nature of the built environment which will ultimately inform their professional decision-making when qualified.

Chapter 8 describes the organization of surveying practice and the position of women within it, giving particular attention to a consideration of the horizontal and vertical distribution, and segmentation, of women within the profession (compare Crompton and Sanderson, 1990). The following chapter (9) looks at 'what it's really like' as to how women are treated and perceived within surveying offices and what their daily professional life actually entails. This chapter includes much detail relating to apparently 'personal' and 'trivial' matters, which, on the contrary, I consider to be the very building blocks of the whole subculture.

The final section considers the implications for the position of women in the built environment. In Chapter 10, surveyors' attitudes towards land use and development policies are investigated, and the surveyor's role in the development process is discussed, in order to illustrate 'how' the surveying subculture might affect 'what is built'. The final chapter (11) seeks to summarise the main observations and issues, to consider the implications for theory, and to make some constructive suggestions for improving the situation for women both as fellow professionals and as inhabitants of the built environment. In conclusion, the implementation of the latter, it is argued, is dependent in many respects on the achievement of the former.

Notes

1 The phrase 'bourgeois feminist' is a neutral term, which may be seen as simply meaning a businesswoman, or a non-radical feminist. From a more separatist perspective it may be interpreted as someone who has 'sold out' to the system. The way it is understood depends on the reader's own ideological viewpoint.

Is more better?

2 The views and observations expressed in this book are my own, and are based on conversations with other women and men in surveying who will remain anonymous. They should not be seen as reflecting the views of the RICS, Royal Town Planning Institute, Architects and Surveyors Institute, my polytechnic department, or any other surveying body.

Chapter two

Conceptual perspectives

The dimensions of the research

This chapter sets the scene with an account of the theoretical dimensions that informed the research and provided the basis of my model, illustrated with relevant ethnographic observations. The model (Figure 1) should not be seen as a mechanistic explanation, although certain bits of it do actually 'work'. I invite the reader to make her own connections both across the columns, and up and down the levels as she reads the book, to help make sense of it all. In order to develop the theoretical basis of this study, which 'horizontally' spanned across both the conceptual areas of gender, class, and space, and the substantive areas of surveying education and practice, I used material from a diverse range of academic realms (Figure 2). I saw three clear levels developing 'vertically', which for the purposes of this research were seen and defined as the **macro** or total societal level, where the all-encompassing concepts of gender, class, and space (and respectively, the related phenomena of patriarchy, capitalism, and the built environment) were considered; the **meso** or intermediate 'group' level, which included the surveying subculture, but also other meso level social groupings, such as the family; and third, the **micro** or individual level of personal interaction, experience, and attitudes – of both women and men. To demonstrate quite 'how' the personal views of individual surveyors (as transmitted via the intermediary of the surveying subculture) affected the profession's perceptions and policies *vis-à-vis* women and the built environment, I seek to show linkages and causal relationships 'up and down' between these levels.

Applicability of feminist theory

Relatively speaking, the 'big' divisions in feminism (Donovan, 1985; Jagger, 1983) proved less important for this research than the variations

Figure 1 The conceptual model

Factors for consideration, and for possible linkage at each level (*with some of the links shown in italics*)

Levels	Class	Gender	Surveying education	Surveying practice	Space
Macro	Feudalism, Capitalism	Patriarchy Feminism	Anti-academic, pro-practical ethos	RICS Professional ethos	Land Estate management
	Power Control	Sex Feelings	Structure of higher education	Technological smokescreen	Bricks and mortar
	Landed classes Landed interests				Landed estates
	Deference	*Women's invisibility*		*'Front'*	
Meso	Upper middle class Professional classes	Surveying tribe Surveying family	The course Student culture Sport	The firm Surveying subculture Sport	The market Property investment Land (cultural) *'What is built'*
	Entrepreneurial fractions	Bourgeois feminists Gender socialisation	Professional socialisation	Bureaucratic/entrepreneurial dimensions	
	Nested hierarchies of power	Development fraternity	Course subuniverses	Status groups Subuniverses	Development process
	False consciousness	*Gatekeeping Exclusion*	*Channelling Tracking*	*Exclusion Closure*	
Micro	The reasonable chap *'People like us'*	Individual niceness and nastiness *'I feel awkward'*	The right type The good student The normal student	The right type The consumer The map-chap	The financial return The site The building Woman as property *'Everyone has a car and plays rugby'*
		'Domestic trivia' The marginal woman 'Paddle your own canoe'	Individual subjects *'Fitting in' 'Working hard'*	*'Getting things done' 'Getting by'*	

Figure 2 Realms of relevance

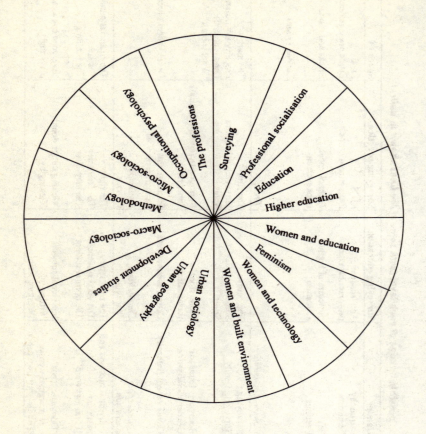

over a much smaller range between the bourgeois and reformist feminists within the middle ground of liberal feminism. These appeared more concerned with 'material' issues (Delphy, 1984; Donovan, 1985: 49–50) and 'practical' feminism, than with developing theories about the forces underpinning society. The concept of the bourgeois feminist became very important, and first hit me when I attended a women surveyors' meeting at which the speaker (not a surveyor) proudly declared of herself, 'I am a capitalist feminist: feminism has been hijacked by the Left', this statement being met with a mixed but not unfavourable reaction. American businesswoman literature, such as work by La Rouche and Ryan (1985), was recommended to me by several women surveyors who were themselves using its principles in the furtherance of their careers. Their philosophy of life appeared to be based on the principle that one should not expect 'special' concessions from the state or your employers. 'You've got to paddle your own canoe', was one of the most frequent phrases used when I asked the women how they coped, which reflected a self-help attitude to life (and sometimes a sense of insecurity, or distrust of society). Hewlett (1988) has brought the message home that American feminism has been more about equal rights within a free market economy, than about socialist feminism, or radical political change. Such matters as the provision of maternity leave (Hewlett, 1988: 298) are apparently not seen as key issues in some branches of the American movement. Many women surveyors are in a sense 'different' from other women because they work in high income areas and so can afford (just) to pay for private solutions to societal problems such as child care. The position of professional women in society (Stacey and Price, 1981: 159) is a complex enigma which combines elements of both comparative privilege and personal self-sacrifice.

Reading widely in the realms of feminist literature helped me develop conceptual linkages at the **macro level**, drawing from a range of perspectives. I studied the various attempts to link class to gender at a macro-sociological level offered by a socialist-feminist perspective (Saffioti, 1978; Hartman, 1981; and Mitchell, 1981). Also the work of radical feminists, who see gender itself as the main causal factor, but seek to study class as a subset of gender (Rowbotham, 1974; Millett, 1985; and Firestone, 1979), was of value in giving a different perspective. Joan Acker's continuing work on gender and stratification from a less radical perspective (e.g. 1973, 1988) was also helpful, particularly in showing the gendered nature of the processes which create class divisions. The differences **between** women was a key issue when dealing with professional women, who had apparently escaped much of the domestication of women in general. There are undoubtedly 'class' differences among women (McDowell, 1986: 313; Walby, 1986), and likewise shared class loyalties between high status women and men (Connell, 1987). Obviously women surveyors are not a unitary group, although on several occasions men surveyors have let slip to me that, 'women, they're all the same really'.

As regards the **meso level**, much feminist literature might be seen as a subset of anthropological research which is concerned with the study of the male subculture by the woman as outsider (Ardener, 1978). Of particular interest is the idea that there are in fact two gender-based cultures in society itself (Hite, 1988: 132) with two separate sets of socialisation processes (Sharpe, 1976: 176; Mead, 1949). I see professional socialisation as a specialised extension of 'male' socialisation. The trick for women surveyors is to 'get it right', that is, to know which socialisation messages apply to them, and to which particular part of their lives, as women and/or as surveyors.

At the **micro level** there is a vast amount of literature which discusses the problems that women encounter at the individual and interpersonal levels at home and in the professional work situation, some of which will be discussed below as part of the substantive material on education and professional socialisation. The Spenders' work on gatekeeping and exclusion of women (Spender and Spender, 1983) was of interest as it complemented and extended the work of Parkin (1979) on closure, from a feminist perspective, whilst demonstrating how it operated at the interpersonal level. Many would argue that men's day-to-day relationships with women are influenced by their sexuality. Popular books that sought to explain 'men' (Brothers, 1981; Ingham, 1984; Hite, 1988), especially those written by men (Korda, 1974; Hodson, 1984), were helpful in understanding the odd and seemingly illogical ways in which some men surveyors treated women surveyors. Metcalf and Humphries' book on the sexuality of men (1985) was of additional interest as one of the contributors was a town planner.

Applicability of mainstream theory

In dealing with a profession, whose role is to serve the landed interests in society, I was naturally interested in material which related property and wealth to class and power in society at the **macro level**. However, since so many 'malestream' theories have been developed without specific reference to women, it should never be assumed that theories developed from observations related to men will 'fit' women, and explain their experience and position in society, without some modification and reformatting. For example, at the macro-sociological level, a range of Marxian literature offered a broad structural explanation of society, and demonstrated that the nature of the built environment might be seen as an end product of complex social and economic processes. Some of the neo-Marxian urban sociological literature was of value in relating class to urban property interests (Simmie, 1981) and in demonstrating the role of the different 'fractions' of the landed professions in the property development process (Dunleavy, 1980; Ambrose, 1986: 68–9). However, I felt a distinct uneasiness with Marx's personal attitudes towards women (which often seem to be held by his modern disciples too), whom he saw as a type of property (Marx, 1981: 222). Likewise, at the other end of the ideological spectrum, Berger and Luckman, whom otherwise I found most helpful in their work on the

Conceptual perspectives

construction of cultural realities at the micro-sociological level, were equally guilty in dismissing 'womentalk' (*sic*) as quite irrelevant (1972: 60). Also at the meso level, material concerning women's representation in the professions was often presented as peripheral, or even discounted as unimportant in traditional studies, no doubt because of the small numbers of women present (Becker *et al.*, 1961: 60; Joseph, 1980: 97).

There is a large group of women who have been left out of traditional class analysis, namely office women, and more broadly the *petite bourgeoisie* of London which includes substantial numbers of female and ethnic groups. I was attracted to works rectifying this neglect (Crompton and Jones, 1984; Crompton and Mann, 1986). Such women possess a tradition of working in offices, banks, and shops (Howe, 1978) and in some cases running their own businesses. Although they are mainly a non-professionalised group, they are of relevance as, like women surveyors, they are working within the male domain of the office situation (Delgado, 1979) with similar 'problems' (Hearn and Parkin, 1987). The boundary between 'professional woman' and 'office lady' tends to blur somewhat in some areas of practice, especially in residential estate agency.

I found it necessary to combine material from traditional 'bourgeois' sociology, concerned with the details of human interaction and social groups, with material from a more radical neo-Marxian tradition which was valuable in giving the broader structural perspective of the urban situation (although some would say that symbolic interactionism can also be built into a theory of society too, Blumer, 1965). Class as well as gender (to varying degrees in different contexts) proved to be a major fact in determining 'who' was considered suitable and 'the right type' within the surveying subculture. However, relatively speaking for the purposes of this study, Weber was of more use than Marx. There is much discussion in Britain as to whether Marxism is only a macro level theory or whether it can accommodate micro levels of human interaction within it too, as evidenced by the debate between the urban sociologists, Saunders and Pickvance. Pickvance's idea of 'nested hierarchies' of power at different levels within society was of interest as a meso level linking concept (alluded to in Pickvance, 1987: 288).

At the **meso level**, I was attracted to the concept of the subcultural group as already mentioned in Chapter 1, and investigated a range of studies of different professional subcultures as indicated in the second part of this chapter. The family, especially the surveying family, also seemed to be a meso level group, although it is all a matter of definition as some would see it as being of macro-sociological importance. Miller and Swanson's work, *The Changing American Parent* (1958), especially Chapter 2, 'Changes in society, and child training in the United States', provided me with one of my most important key themes, which tied in with the concept of the bourgeois feminist. Miller and Swanson, in their study of the upbringing of children in the 1950s in North America, make the interesting distinction between what they call entrepreneurial and bureaucratic cultures of different middle-class families, that is, respectively those families which are part of the business community and those which are oriented towards public service in government agencies. Both boys and girls brought up under the former influence are likely to go into the private sector and possibly have their own businesses. Although intelligent, they and their parents may feel somewhat ill at ease with the requirements of academic education. Those in the bureaucratic category are more oriented towards public and governmental service, exhibit all the traits that are rewarded by the school system, and may be more likely to conform to traditional gender roles. Miller and Swanson's work puts considerable emphasis on girls as well as boys. It is full of valuable observations, which I saw echoed in the lives of women surveyors. For example, entrepreneurial mothers may be keen to have their children cared for by childminders and nannies: and far from causing maternal deprivation, they see it as a means of developing independence in their children. More recent work from the sociology of education on the different fractions within the middle classes echo some of these ideas (Bernstein, 1975; Davies, 1976: 129: Delamont, 1976).

I see the **micro level** as the operational level where people's judgement of each other and their suitabilities determines who gains entrance and progresses within the subcultural group. The ideas of Berger (1975: 128) on person selection of those that 'fit' and of Merton (1952: 361–71) on the bureaucratic personality were of interest. Although women surveyors seem to be more entrepreneurial in personality, as will be explained in later chapters, men surveyors can adopt bureaucratic or entrepreneurial demeanours depending on the societal and organisational situations in which they find themselves. One cannot generalise too much, or create foolproof conceptual 'types', as the situation is extremely complex. Indeed, it seemed to me that bureacratic and entrepreneurial types were flip sides of the same patriarchal coin. Katie Stewart's work (1981) on male bonding and ranking cast further light on this apparent paradox in showing that men can make alliances with other men who may hold opposite views from themselves, but derive from similar social backgrounds.

Broader spatial perspectives

I was especially interested in literature which made the linkage between society and space. It was helpful to see 'space' as including land, the built environment, and landed property interests. As stated, a neo-

Marxian perspective was, to some extent, of value in demonstrating the inter-connections between social relations and space (Dunleavy, 1980; Massey, 1984). The relationship between social processes and the spatial end product could also be facilitated by other less 'threatening' theories such as urban systems theory (McLoughlin, 1969) which was remarkable for being all encompassing and yet apolitical (Simmie, 1974; Bailey, 1975). Such theories were devoid of any gender dimension, but were of value in helping to see 'the whole'.

Gender does not exist in a vacuum, any more than does class, and both must leave their imprint on space, that is, the built environment. The study of the interrelationship between gender and space, and the question of whether this 'new' duo can be related to existing theories concerned with the interrelationship between class and space – involving a major reappraisal of urban conceptualisation – has exercised the minds of feminist geographers over the last few years leading to major debates in journals such as *Antipode*, a radical geography journal (Foord and Gregson, 1986; McDowell, 1986). There has also been a growing amount of substantive material giving a gender perspective to familiar urban issues and problems (Greed, 1987a) – in the process redefining the urban agenda (firstly from North America, Torre, 1977; Wekerle *et al.*, 1980; Keller, 1981; Stimpson *et al.*, 1981; Hayden, 1984; and then from Britain, *Built Environment*, 1984; Matrix, 1984; WGSG, 1984; Cockburn, 1985a; GLC, 1986; Little *et al.*, 1988).

Space has other more mystical meanings. Cockburn in her book on male domination through technology quotes Connell, a male sociologist who has made a study of gender, as stating that to be male is to occupy space (Cockburn, 1985b: 213). The landed professions by their very nature reflect this particular trait of masculinity. This ethos directly affects the surveyors' perception of women's natural 'place' within the built environment *and* their role as professionals. Taking the importance of male control further (of which domination over land is but one manifestation), French (1985) sees 'control' (power) as a primary causal factor, almost in a feminist interpretation of Weber. Nature is seen as the main force that has to be controlled, and women and land are seen as part of nature (Griffin, 1984). Other elements on the agenda of the landed professions that have to be controlled are the future (Cross *et al.*, 1974), the working class and their housing, and the market. The role of surveyors as controllers, managers, and planners, as well as businessmen and professional gentlemen, comes out strongly in the historical account of the development of the profession.

Women have been seen as 'space', that is as property themselves, both as wives (Holcombe, 1983) and as virtual chattels in rural estates where they might be seen as mere figures in the landscape (Dresser, 1978). Not only were women property, most could not own property

(Hirschon, 1984) until relatively recently. Women have entered the modern period without their relationship with real property being satisfactorily resolved within English law in respect of divorce and intestacy (Freund, 1978; Midland Bank Trust Company v. Green, 1979). This historical background still influences the way in which the landed professions view women today.

Can all this be observed?

Whether investigating the macro, meso, or micro aspects of the situation it was important always to look behind apparent realities and 'facts' and consider the cultural perceptions that created the images of reality (as seen variously by the researcher and the researched). There was much helpful material to assist me in my task within some branches of the social sciences (Goffman, 1969) including, more recently, the more sociological areas of geography, where, for example, Sayer's work on realism (1983) has aroused considerable interest. Sayer's idea of 'unpacking' meanings and looking behind the facts as they appear was particularly appealing. Also, it was of great value to meet people from other academic realms at conferences, who did not share the values of the surveying world, which helped to 'make the familiar strange' (Delamont, 1985).

As regards the model, people do not carry around in their pockets a blueprint of capitalism or patriarchy to which they refer in their daily lives for guidance (Greed, 1987b). Rather, they go about their business doing what comes naturally and what makes sense to them. What is amazing is that so often they seem to be unwittingly fulfilling the requirements of various theories in the process as agents/actors (Giddens, 1984). As a researcher in the heat of the moment, I find it hard to carry in my head an encyclopaedia of all the theories that might be relevant in a specific situation. Real life is not simple and one cannot observe a surveyor's attitudes and separate the various strands as being caused by patriarchy, capitalism, or whatever into neat theoretical compartments, nor can one always 'see' inequality as one may have to wait for certain occurrences to show their true fruit in the course of time (many women begin to only see what is really happening in their careers and lives in their thirties). However, one can build up a generalised picture of sets of attitudes held by surveyors towards class and/or gender factors (and it's surprising how often the two go together in the same observation). For example, among certain types of men surveyors, a veneration for land is often accompanied by a deference for the landed classes; a rather traditional conservative political viewpoint, a love of getting out and about, either on site or in the country, a strong group identity and love of team sport, but a somewhat limited understanding

Conceptual perspectives

of 'social' as against 'real' professional and business matters, and a *gauche* or over-*gallante* attitude to women. Such attitudes inevitably work their way across 'onto space' via the vehicle of professional decision-making, resulting in emphasis being put upon the needs of the car-owning, sport-loving, affluent male – often at the expense of other interest groups within society.

The sociology of education and the professions

Some reservations

The purpose of the second half of this chapter is, first, to identify material from these realms which contributed to the development of the conceptual levels, and second, to discuss substantive material of direct relevance to surveying education and practice that provided insights and 'sensitising concepts' (Hammersley and Atkinson, 1983: 179). Like the urban geographers, the educational sociologists (both traditional and feminist) have difficulty dealing with class and gender together, let alone the trio of class, gender, and space. In the mainstream literature of the past, if women appear they are more likely to be seen in stereotypical supporting roles as 'the mother', rather than as pupils or students themselves. In reviewing the literature of the sociology of education one is confronted with the problem that, until relatively recently, much of the emphasis appeared to be on studies of male working-class boys (Acker, 1983a) within state schools (Willis, 1977), who typically are in conflict (Cohen, 1976) with the educational system, or are seen as 'problems'. In contrast, women and men surveying students may be seen as those that have succeeded, and are at college because they want to be there, and they themselves are from relatively privileged, problem-free backgrounds.

Whilst the balance is nowadays being corrected by the tremendous growth in gender-related research, I had considerable difficulty believing some of the feminist literature that purported to describe the experience of 'all' women. Many feminist books make much of how girls are socialised in school (Deem, 1980; Thompson, 1983: 37; Stanworth, 1984) into accepting the primary importance of having children and becoming housewives, thus cutting down their career horizons. This was not what many women surveyors had experienced. Many women surveyors told me that in their schools they were constantly encouraged to have careers, indeed to such an extent that anyone that wanted children was likely to be seen as rather 'thick' and would be likely to be in the non 'O' level stream studying cookery or art.

Women surveyors told me they were more likely to be encouraged to be doctors, lawyers, or accountants than surveyors, because of the perceived 'roughness' of surveying (the exact opposite of the reality) (compare, Elston, 1980: 111). Sharpe comments that bricklaying is seen as unladylike (1976: 176) and by association it would seem that this attitude transfers to the landed professions in general. It is fascinating that some girls are told they cannot do surveying because it is seen as too scientific, but paradoxically a career in medicine is encouraged. In contrast, Mahoney (1985) notes that boys are likely to receive talks on surveying in which it is portrayed as an acceptable profession.

Of course, the class factor must be taken into full account, as many women surveyors went to girls' public schools or 'selective' schools, or at least were of a higher social class. The literature on public schools has traditionally been male-oriented; so the school background of women surveyors and other upper-middle-class women going into professional areas seems to be missed out, falling down a gap between studies of gender or of class. However, there were some valuable exceptions of studies of girls in public schools (Delamont, 1976, and Okely, 1978; and also Delamont, 1984; Atkinson, 1985: 161, and Walford, 1986).

Conceptual levels

I will now look at material, from both the literature on education and of the professions, of relevance to my three levels of macro, micro, and meso. Education has a major role to play in the development of the nature of society, and there is much literature at the **macro level** on its role in the reproduction of social relations and transmission of the values of the 'ruling classes' (Bernstein, 1975; Bowles and Gintis, 1976; Sarup, 1978). Material that dealt with the role of education in controlling and limiting access to the professional classes and thus maintaining the class system was of great interest. The concept of credentialisation (which derives from North America), that is, the limiting of access to specialised professional enclaves by means of increasing the educational entrance requirements (Collins, 1979: 90–1) seemed, to some extent, applicable to the development of the surveying profession, as will be described in the historical account.

One must be cautious in using American literature, as Britain is unique in having what appears to be relatively easy access to the higher professions by simply completing undergraduate courses or undertaking 'articles', without the need to attend expensive, difficult, and time-consuming graduate school. There are many additional ancient class and status factors involved which limit the progress of those who are 'not the right type'. Credentialisation does not always seem to be a valid theory in a profession such as surveying in which 'it's not so much what you know as who you know that matters'. Many women commented to me that every time they work out and achieve what is required in terms of qualifications or experience, 'the men move the goal posts and change the rules'. Credentialisation might be used to legitimate the *status quo* as I came across several people getting further qualifications, after they had achieved professional status.

I was intrigued by the work of Olin Wright (1985), who convincingly links bourgeois sociological concepts deriving from traditional studies of professional groups, and credentialisation in education, with macro-sociological concepts from a neo-Marxian perspective whilst retaining a gender perspective of sorts. Thus he creates a workable, if somewhat confusing, overview of society which he grounds in the concept of exploitation as the basis of class analysis. Whilst having reservations with his theoretical basis, the sheer sweep of his conceptual framework gave me further inspiration for the development of my theoretical model.

Other theories of education at the national level almost contradict ideas of credentialisation, for example, models of educational expansion based on economic 'man'power theory and national resources which seem more applicable at times of economic growth (Blaug, 1972, Woodhall, 1972: Chapter IV, 'The demand for educated manpower'). One must always look at what the true outcome is for women in a period of either educational growth or 'man'power shortages, or sheer educational 'inflation' (Dore, 1976). If women are 'encouraged', one has to consider what role and position they are likely to hold within the professional organisation. Increased numbers in themselves mean very little: women make up half the pupils in schools but that in no sense ensures either equality in education or improved career advancement in work, thus the outcome of broadened entry is crucial. As Joseph noted (1980), academic achievement and qualifications are not concomitantly linked to progress within the profession (even for men) – quite the opposite in some cases. Indeed this proved to be one of the major complaints of the women interviewed. However, men, now, appear keen to recruit more women to keep the numbers up (*Estates Times*, 10.3.89, No. 985: 1, 'RICS Educational Shake-up'[1]), particularly because of the 'demographic time-bomb' factor, but whether this is to the benefit of women is another matter.

The concept of professional and business elites has been important in 'bourgeois' sociological conceptualisation of power structures. I investigated theories on the role of the professions within elites (Mills, 1959; Bottomore, 1973) and the personalities they engendered (Gerth and Mills, 1954). There seemed to be less substantive, rather than theoretical, material from neo-Marxian writers but the work of Westergaard and Resler (1978) was of interest. A helpful summary of the state of the art of the sociology of the professions is an article by Saks which classifies and clarifies the various approaches (1983).

The role of women in elite professional groups is unclear. They may be seen in a supporting role as wives (who must look 'right' as in Whyte, 1963: 292) or as the 'same' as the men but twice as 'guilty' (Callaway, 1987). The literature is full of contradictions as regards women (Silverstone and Ward, 1980). For example, female elite professional groups may merely be seen as quasi-professions (Etzioni, 1969), even when they do exert considerable power (if 'only' over other women or in welfare agencies). Even when women belong to an elite group in terms of family and class, they may still occupy the role of a sub-proletariat within that group, and may have very different expectations, and simply 'be treated differently in the same situation' (Benn, 1979; MacDonald, 1981). Indeed, the privilege of an expensive education may actually hobble them for life (Okely, 1978). However, other studies suggest that powerful female elites can co-exist and succeed alongside male elites from the same families (Delamont, 1976). In contrast, some would argue that individual women from non-establishment, shopkeeper, marginal, and so-called 'lower class' backgrounds are likely to do better (because of less mainstream conditioning), provided their family has a business orientation of self-employment, and a mentality of risk-taking independence (Hertz, 1986).

The material on the **meso level** proved one of the most useful areas, in terms of the extensive literature on professional and educational subcultures, and on the use of ethnographic methods in the study of these subcultures. The question of whether the student subculture was positive or negative in function was of particular interest in examining the relationship between the educational and professional subcultures in surveying. As Joseph points out, Willis (1977) identifies an anti-academic, pro-practical subculture amongst his working-class boys which was assumed to be 'bad' by their teachers; *but* it is seen as 'good' when similar traits are exhibited by surveying students, as will be explained.

It was important to look for material that linked home and personal life to the school or college, as many women surveyors considered their background and present family commitments as key determinants in both educational and career progress. Although there were some books that did this admirably (such as Miller and Swanson, 1958; Douglas, 1967; Connell *et al.*, 1982) I was still left with an overall feeling that all life ceased after one left the school gates. Indeed, it is a characteristic of much malestream literature to compartmentalise life and deal with 'work' situations, including education, in isolation. It does not deal with those matters which are of no concern to men whilst they are at work, such as child care or domestic duties, because presumably they take it for granted that their wives are doing this for them. More negatively, in traditional literature some home/work linkages were made, but they

Conceptual perspectives

were more concerned with the level of support the wife gave to her husband's career (Pahl and Pahl, 1971), than with seeing the mother and the student, or the wife and the professional, as one and the same person as in more gender-related literature (such as Rapoport and Rapoport, 1971; Finch, 1983; and Lorber, 1984: Chapter VI).

At the **micro level** of interpersonal relationships and classroom interaction, there is now a substantial literature related to women's experiences of school education. Whilst there is much material regarding schools, there is nothing like the same amount of material concerning women's experience in higher education. Do all those awful things that some of the feminists go on about regarding mixed state schools happen in polytechnics too? (Loban, 1978; Deem, 1984; Jones, 1985; Mahoney, 1985). Higher education is also mixed, but there are many other factors to take into account especially class, selection, and motivation. There are definitely some 'problems', but they manifest themselves in a more subtle way with surveying students; and as several women commented, 'there is simply no point in making a fuss'. However some of the literature rang true for some of the women in surveying. As one of my sanest women ex-students said, 'there was a stage when I thought I was going mad, we were always told everything is equal now, so I thought it was just me. It wasn't until I got out in practice that I met others who said exactly the same things had happened to them too'.

I was interested in developing linkages between the conceptual levels, and was attracted to a range of theories which tried to explain the role of the subcultural group at the meso level as a transmitter and enforcer of social control for those in power in society above at the macro level, through the attitudes and actions of the individuals in the group. I studied the ideas and diagrams of Bourdieu (1973), Bernstein (1975), Delamont (1976), Atkinson (1985), and Olin Wright (1985), all of whom in their various ways make linkages between the macro level to the micro through the meso level, retaining a broad structural view of society without getting bogged down in subcultural detail for its own sake.

Surveying education

I will now discuss substantive material of direct relevance to the areas of surveying education and practice. As will be explained, in a sense, the whole surveying subculture is out of step with much of academia, because there is a strong practical as against academic emphasis which tends to encourage both lecturers (Stapleton and Netting, 1986) and students to decry 'theory' or learning for its own sake. Davies (1976) also notes that a rather laid-back attitude towards academic study is not uncommon among high status public school boys (although they might be working like mad on the quiet!). Some women students also subscribe to this anti-academic attitude, although for women 'it's different' as they perceive that they are judged by a double standard and expected to work twice as hard, or they will be told that they are not really interested.

I sought material to explain the actual setting of surveying education and the place of women students and lecturers in academia. I frequently got the distinct feeling that the 'normal student' and 'ideal lecturer' were male and that women were still a special category. There were exceptions such as Whitburn's study *People in Polytechnics* (1976) which dealt quite unselfconsciously with women as well as men. However, much of the foundational literature on polytechnics, where around 80 per cent of surveying courses are located, was alarmingly gender-neutral (Robbins Report, 1963; DES, 1966; Robinson, 1968). Overall, the role of women as academics is open to much debate (Acker, 1983b; Cass, 1983; Acker, 1984).

Feminist material concerning the nature of the subjects taught (Culley and Portuges, 1985; Whyld, 1983) especially in the landed professions (Weisman and Birkby, 1983); how they are taught (Weiner, 1985; Bunch and Pollock, 1983); and the educational organisations and spatial settings in which education is carried out (Langland and Gove, 1981; Stanworth, 1984), were all of value in developing sensitising concepts (Hammersley and Atkinson, 1983: 179).

Joseph's findings

At this point it would seem appropriate to introduce in more detail existing work on the subculture of surveying education. Apart from Joseph (1978; 1980; 1988: Chapter 16), there is the occasional, but rare, student who has written about surveying education himself (for example, Davies, 1972; Wareing, 1986). Indeed, Wareing found that only 1 per cent of his student colleagues would want to go into surveying education, even if they were offered head of department status! (In fact only 1 per cent of men and 2 per cent of women surveyors are in surveying education.)

I found Joseph's research invaluable in confirming many of the feelings I already had about surveying, and fully acknowledge the value of his work, his encouraging interest in this study, and thank him for his permission to reproduce his findings here. However, with respect, I had some misgivings about his methodology, and of course the inevitable lack of women. There were three elements which Joseph investigates in his study: (i) the values of the surveying profession, (ii) the question of whether students were socialised to surveying before or during the course, and (iii) the nature of the professional culture. First, as a result of studying surveyors whilst in college and then following on with a short longitudinal study, he concludes that those that reflect the 'correct' professional values, with the greatest level of 'precision' (as tested by questionnaires) are the most likely to succeed, and that this is not the same as success in examinations. Rather, it is a matter of those that most strongly reflect the values of the surveying subculture and who are likely to have family connections that are most likely to succeed. The second question was concerned with whether these values were 'confirmed' rather than 'conferred' as a result of taking the surveying course, He observed 'anticipatory socialisation of prospective students', that is, Joseph sees students turning up for the course with their values preformed.

Joseph refers to the work of Becker *et al.* (1961) who stressed the idea of the development of a 'perspective' acquired by an ongoing interactive socialisation process with the subculture as the end product, rather than seeing the subculture as a 'fixed' institutionalist element that students are expected to internalise. Therefore he questions Becker who sees the professional socialisation process as a major aspect of professional courses. It would certainly seem from my research also that for men, and especially for women, the most successful surveyors are those that come from families which are already part of the professional subculture, indeed I came across veritable surveying dynasties. The women who can 'handle it' best are those who are already cognisant of what to expect from men, that is, they are having their suspicions 'confirmed' (Greed, 1987c) by professional education, and they can cope when sexist images are 'conferred' upon them (in a kind of reverse or negative socialisation process, reserved for women).

Joseph uses the word 'culture' frequently, and indeed, significantly states that Gerth and Mills (1954: xxii) say it is one of the spongiest words in social science. In the third part of his study, he defines the 'culture of surveying' by identifying what surveyors see as 'good words' and 'dirty words'. As can be seen, subjects such as planning and economics have the lowest status, and are seen as waffly subjects. Law, valuation, and construction and land survey are good subjects, because they are practical.

Dirty words: academic, theoretical, planning, economics, and education.
Good words: practical, relevant, sensible, law, land, surveying.

He identifies four key concepts, which were the basis of his questionnaire to test the correctness of the values of his students. These were practicality, success, professionalism, and freedom which he defined as follows: surveyors are motivated by a desire for 'practicality' which means dealing with 'real' issues and a desire 'to get the job done'. He notes a suspicion of anything sociological or abstract or indeed learning for its own sake (however, many surveying students have good 'A' levels). 'Success' is measured in the esteem of peers and in professional 'status' rather than in monetary terms. Professionalism is all tied up with a sense of duty and service to the client and also a strong sense of veneration for 'land', almost in the sense of serving it, as if it and not the human were the client. Freedom is expressed in a desire to have the independence and control over one's time that a professional partnership or at least one's own business gives, and again relates to the freedom of 'not being stuck in an office all day but getting out and about'. (I remind the students of this in cold weather when we go out on site visits.)

Some of this orientation has changed radically within the last ten years as the ethos of surveying has become more commercial and unashamedly concerned with financial success within the enterprise culture of the 1980s (Greed, 1988). Whilst 99 per cent of all surveying students have always got jobs, at the time of Joseph's study over a third of students were finding jobs in public service, whereas nowadays the figure is nearer 10 per cent and indeed last year around 80 per cent of my college's female estate management students found jobs in prestigious London practices. The relative status of some of the subjects studied has shifted. The construction and land surveying component is now seen as more down-market. A whole new range of 'practical' market-oriented subjects, including computers, modern valuation techniques, and even management and business studies, have taken their place in importance, and economics is no longer a dirty word.

The big question is how do women students fare within this subculture; indeed, do they possess a second subculture of their own? Some educational studies of female subcultures related to working-class situations note that 'conflict' manifests itself in silence and sullenness rather than outright aggression. In the case of the females in surveying, it is very different in that they want to be there and they are, in a sense, twice as motivated as the males. However, because it is still considered odd in some circles for females to be surveyors they may be seen as 'deviant' or in conflict with society. I have come across several examples of females being told by their schoolteachers that they were being awkward, or naughty because they wanted to become surveyors, and being virtually punished by women teachers for it (Wigfall, 1980; and compare Whyte *et al.*, 1985). Paradoxically, women surveying students are often ultra-conservative in their views and so here it is a matter of inter-class misunderstandings between women. However, this reflects a deeper issue that may owe its existence to the power of patriarchy ineffectively administered, which is the fact that women receive a variety of conflicting messages as to how they should be

Conceptual perspectives

socialised, as women and as surveyors. Indeed, many women have told me how difficult it is to be a woman and a surveyor at the same time. However, it would seem that women from entrepreneurial backgrounds are more likely to take both in their stride (Miller and Swanson, 1958).

The surveying profession

There is a range of standard texts on the professions which mention surveyors and their 'unusual' characteristics briefly in passing. For example, Millerson (1964: 141) states 'thus qualifying associations benefit through these late developers seeking qualification', with reference to those surveyors who took their professional examinations in their early thirties. Indeed, the surveyor's measure of personal development and success is not necessarily based on academic qualifications and in the past many successful and competent professional men took many years getting round to becoming completely qualified. Saunders and Wilson (1933: 194) describe surveyors as 'an unusually wide variety of specialists'. Hurd (1978: 134) remarks that Lewis and Maude's work (1953) seems dominated by a spirit of nostalgia, and I had difficulty seeing my women surveyors fitting into the 'tobacco-laden' image of the professions created by these books.

Material related specifically to the ideologies and ethos of the landed professions was more useful (Knox, 1988) and some of it was even written by women and/or included a gender perspective (Wigfall, 1980; Estler *et al.*, 1985; Howe, 1980; Howe and Kaufman, 1981; Nadin and Jones, 1990). The latter was of relevance for comparative purposes, in demonstrating that town planners' professional decisions were influenced by their personal views. Of particular interest were articles on the nature of the town planning tribe (Marcus, 1971; Walsh and Gibson, 1985; and RTPI, 1984). In contrast, little attention has been given by academia to the study of the professional work of surveyors, with the exception of an article by Dickson on the management of surveying practices (1985), and of greater value for my purpose, the work of Martin Avis and Ginny Gibson (1987) on the management structures of general practice firms.

A wider perspective was gained from material on the growth in the numbers of young women choosing professional careers (Silverstone and Ward, 1980; Gerstein *et al.*, 1988); and on the situation where they formed a minority in professional education (Kleinman, 1987; Thomas, 1986: Dewar, 1987); and in practice, such as in law (Smart, 1984; Spencer and Podmore, 1987), medicine (Elston, 1980; Lorber, 1984), business management (Kanter, 1977) and science (Burke, 1985; Irving and Martin, 1985). Dorothy Griffiths' work on women in technology was of great relevance, especially her observation that technology is even more male than science (1985: 66).

A major report by women solicitors on the position of women in the legal profession (Law Society, 1988) gave practical suggestions on how it might be organised to fully take into account women's 'other' role. Solicitors are much further ahead as 50 per cent of new entrants are now women, but much further behind in some respects as law is generally considered to be one of the most patriarchal professions regarding content and practice. It is always interesting to check on material on women in other professions, to see what 'the going rate' is compared with the surveying profession. Such literature casts light on the mechanisms that were at work in filtering and channelling women horizontally and vertically within the various professional organisations. Longitudinal studies such as Woodward's (1973) that linked women's educational experience to their subsequent role in practice in male-dominated areas were fascinating, but rare because of the timespan required. Also, from a more negative perspective, there have been several studies of what can only be described as the hatred of women exhibited by some professions, especially medicine (Oakley, 1980; Savage, 1986). Surveyors are generally fairly chivalrous and gentlemanly towards women in comparison, but as will be shown, some women consider that this attitude can be equally negative in the long run. The Women and Housing Working Party Report (Levison and Atkins, 1987) was most illuminating, showing the lack of progress of women in spite of the apparently 'girl-friendly' (Whyte *et al.*, 1985) ethos of housing.

I consider that the micro level 'nicenesses and nastinesses' of everyday interaction between individuals provide the building blocks of the whole edifice of the surveying subcultural group. This led me into an investigation of relevant areas such as management studies, and what may broadly be described as occupational psychology. It is noticeable that there was often a resounding silence about gender in mainstream work (Hearn and Parkin, 1987: 4). However, the situation is changing, for example, Kanter had written in the 1970s about women and men in large organisations, and by the 1980s her work on interpersonal relations had penetrated the mainstream market too (Kanter, 1977; 1984).

Books which comprise studies and/or advice to career women such as those by Gallese (1987), Cooper and Davidson (1982), Marshall (1984), Coote (1979), Hennig and Jardim (1978), and Williams (1977) were very helpful. I was also very taken by what might be seen as some of the more 'pop' material on management such as CareerTrack's conferences and publications (White, 1987) which are attractive to some women surveyors in apparently giving them the key as to how to

Conceptual perspectives

succeed (without any major change being required on the part of the men). I was fascinated by concepts such as power dressing which seemed to blame the inequalities of the world on the fact that women wore pink instead of navy blue, or that they spoke with too high-pitched a voice, or shook hands wrongly, or were not assertive enough (Dickson, 1982). Other women surveyors, especially older ones, seemed mildy amused and unimpressed by such material. Some men surveyors break all the rules on personal presentation, and are still seen as the right type, so all this seemed very one-sided to me. As one woman put it, 'behind the rugby changing-room doors, that's one place we can never go, heaven knows what they get up to in there'. The world (and men) had not changed that much for women in spite of them doing all the things the books told them. I concluded there must be some additional Factor X, which I never identified, which would explain it all.

Note

1 References to the *Chartered Surveyor*, the journal of the RICS, and to the *Estates Times*, a popular weekly publication read by a wide range of property professionals, do not have a stated author and may simply be notices, obituaries, reviews, etc.

Part two
The historical perspective

Chapter three

The background to surveying up to 1900

The origins of surveying

Thompson, in his invaluable history of the surveying profession, observes that the skills of surveyors have always been needed in 'orderly and property conscious societies' (Thompson, 1968: 1). To unpack this statement, it is clear that surveyors have seen themselves, not only as high status professionals but as custodians of civilisation itself. This attitude is echoed in recent times in an advertisement in the RICS journal which states that property is the basis of civilisation and that art is one of its great achievements (*Chartered Surveyor*, 3.10.85, Vol. 13, No. 1: 9 of Supplement). Surveying is the profession concerned with **ownership**, dealing with the valuation, auctioneering, and management of both real, and personal property, e.g. livestock, houses, estates, and even shooting rights, and antiques and machinery, and nowadays office blocks, retail centres, and investment portfolios. The ethos of surveying strongly espouses traditional conservative values, and applauds possessive individualism (MacFarlane, 1978).

Historically, surveying developed to meet the needs of landed wealth which owed its origins to feudalism (Fitzherbert, 1523). As late as 1875 surveyors were advising the owners of over three-fifths of the landed property in England (Thompson, 1968: 167), the bulk of which belonged to a very small number of families. Over two-thirds of the land throughout the centuries was held as very large estates generally as 'settled land', where dynastic patrilinear ownership rather than individualistic owner occupation was the rule (Thompson, 1963). Early surveyors were more concerned with land management (Leybourn, 1653), rather than property investment and transfer, as most land was inherited rather than bought. Trade in land as a commodity did not come until later with the growth of capitalism. It is interesting that even today surveyors still see themselves as primarily estate managers (and not estate agents). However, in the modern commercial world, this emphasis on 'management' has an added twist, being directly linked to the property market and the wider business world, rather than to the needs of individual owners. As Enid Harwood put it (the first and only woman president of any surveying body namely the Faculty of Architects and Surveyors, in 1987–88, Appendix 2), 'a growing commercialism is blowing away the restrictive practices of a once privileged age' (Harwood, 1987). Nevertheless, the historical expertise in land (Joseph, 1980) is still stressed to legitimate the surveyors' self-image and right to a special professional monopoly (compare Goffman, 1969).

Over the centuries, a fine distinction developed between the work of surveyors and accountants who were also concerned with wealth management. Even when landed property was converted into money in the modern era, surveyors retained control of investment related specifically to land. Surveyors also had to resolve their territory with 'the monstrous regiment of lawyers' (Thompson, 1968: 29) who made several attempts to take over the pitch (compare *Estates Times*, 14.8.87, No. 907: 36, 'RICS . . . war against solicitors'). One has to be aware of these ground rules and territories amongst men, before even considering women.

From a class perspective it should be remembered that only 10 per cent of all property was owner-occupied before 1914 (Merrett, 1979: 1). Some would argue that, even today 90 per cent of the wealth is owned by 10 per cent of the population (Norton-Taylor, 1982), and that the real power in Britain is still feudal rather than capitalist, or perhaps the two have merged. Whilst one could argue that nowadays the really big landowners are financial institutions (CIS, 1983) and that we live in a property-owning democracy in which over 60 per cent of housing is owner-occupied; it is important to remember that (incredibly) only 10 per cent of Great Britain's land surface is urbanised (Best and Anderson, 1984: 22), and that big 'feudal' estates still cover much of the space in between. Space matters in measuring power (Ardrey, 1967; Ardener, 1981: 26). Whether feudalism which is based on land, or capitalism which is based on production, is the more patriarchal is a question of great interest when dealing with the position of women in the landed professions, 'we always take the best man for the job, even if she's a woman . . . women are better value for money'.

The fact that women now have 'the same' opportunity as men to be surveyors, to serve this landed elite with its strong patriarchal heritage, but not to change it, is a strange form of equality indeed. Both women surveyors and the modern mortgageriat are nowadays joining a very ancient club that was created by, and for, very different types of people. Even if women have the right to enter, one still comes across a certain male huffiness and 'lack of respect' that no doubt is a reflection of the effect the past has had on shaping men's values. I well remember leaving a meeting of women surveyors early, at the RICS headquarters, feeling elated, when I happened to overhear a couple of men surveyors asking the porter, 'what are all these women doing here?'

Until relatively late in history, women were virtually seen as property themselves. Thompson (1968: 16) writes of 'highly active land and marriage markets' in Elizabethan times. By comparison, several women have stressed to me the importance of their possessing negotiating skills similar to those required by the traditional woman marriage broker, seeing this as a positive attribute rather than a sexist factor (compare Hillel, 1984).

In the historical rural estate, women were likely to be seen as either members of the owner's family, or as estate workers, but not as potential surveyors. I came across a book written by a woman rural surveyor who was a land agent on a large rural estate in the 1950s, but one is given the impression that so little had changed it might have been the 1590s (Napier, 1959). The book is a sensitive, retrospective, sociological account of the author's experiences (although camouflaged in a humorous style). She graphically describes the tenants' reaction to her, which reflected centuries of unwritten rules. They did not know what to make of her and one of the estate workers bemoans that 'young ladies was young ladies when I was a lad'. They tried to drop her, accidentally on purpose, into the slurry pit, to put her in her place. Read from a modern feminist perspective, one could see the whole book as an example of every sort of harassment imaginable. However, she was exceedingly resilient and took it all in her stride, expecting no 'special' treatment – a characteristic of many of the 'first women' of the time (and today).

One cannot generalise too much about the past, as some women did own property themselves. The inventory of property interests after the Great Fire of London (Mills and Oliver, 1967) suggests considerable female representation. Indeed, as the centuries rolled on, the situation actually deteriorated, until the reforms of the late nineteenth century (Atkins and Hoggett, 1984: Chapter 1).

Changing status

Gradually certain surveyors broke away from the landed estates and set up in private practice. They were likely to have come from an agricultural background themselves, and met the professional needs of yeomen farmers on a freelance basis. For instance, in 1760 John Player, a Gloucestershire farmer, extended his surveying activities to a full-time practice (Sturge, 1986: 4) later being joined by his nephew Jacob Player Sturge. Some of the most prestigious London firms of today are also derived from relatively humble origins. For example, in 1765 William Clutton, the third son of a vicar, married his employer's daughter and

The historical perspective

thus took over an existing small surveying practice (Cluttons, 1987). Those that were started by relatively high-class founders seem to have kept that elite status right up to the present day. For example, Edward Strutt, a founder of Strutt and Parker, was a yeoman farmer himself (Strutt and Parker, 1985: 1) and fifth son of a lord, and had attended Cambridge. The diverse social origins of the early surveyors were to be replicated in the 'type' of young men that were attracted to surveying for many years, and to some extent even today. They included younger sons of landed families, keen young men from non-Establishment backgrounds, and 'perfectly ordinary chaps' with a tradition of family business or farming backgrounds, plus various socially 'marginal' and bright men.

Women had some involvement in the professional activities of these firms. For example, Maria Savill (1807–94) took over the family firm which at that time included a building business (Watson, 1977: 81). A more recent example is Margery Dawes, who virtually ran the family firm in the early part of this century and was known as a formidable figure at Hoddell Pritchard, a small provincial practice (Hoddell, 1985). Likewise the wives, in the prestigious firm of Drivers Jonas, seemed to be expected to know the business as explained in Barty-King's history of that firm (1975). It would seem that a wife would need the equivalent of a surveying degree to converse with her husband. For example, the love letters of a young surveyor to his sweetheart in 1803, included in Barty-King's book (1975: 49), are an incredible mixture of passion and property. The private and the professional realms had not yet been divided in this family firm. No doubt such women nowadays would be more likely to qualify as surveyors themselves, and indeed they do.

Thompson notes the importance of large surveying families that formed dynasties, and produced the elites within the profession (1968: 234). It is noticeable that many of the male elite group, from which presidents of the RICS and senior partners of the prestigious firms are drawn, are likely to be married to daughters of surveyors (e.g. *Chartered Surveyor*, January 1947, Vol. XXVI, Part VII: 415). One could argue that women have always had a prominent position in surveying, the difference being that nowadays they have formal qualifications. Interestingly, women surveyors today, although a small minority are more likely than men (proportionately speaking) to be found in the high prestige elite practices, and are often seen as being 'classier' than equivalent men in practice. Whether their role is the 'same' as that of the men (Delamont, 1976) or a re-enactment of the historical role of the helpmeet (Heine, 1987) or even that of a micro-proletariat is another matter (MacDonald, 1981). It would seem that women are more likely to achieve low status positions in high status practices, than high status positions in average practices.

There were no qualifications or exclusionary mechanisms based on education at all until the mid-nineteenth century; rather, surveying was a skill passed on from father to son, the ultimate exclusionary mechanism for women. The status of a particular surveyor related to the value and type of property he dealt with, and the opinion of his clients. Some individuals were small tradesmen who dealt with property transactions, and might also have done a spot of auctioneering and combined this with being the local coal merchant or undertaker too. For example, Alonzo Dawes, one of the founders of a small provincial practice, was originally an auctioneer, valuer, and coal merchant (Hoddell Pritchard, 1985: 3). At the other extreme there were high status and rather risky auctions 'by the candle' in Garraways Coffee House, Cornhill in London, for wines, and later lands and houses (Thompson, 1968: 48), organised by 'posher', but not more qualified, gentlemen. Already the important London versus provinces division was becoming visible.

There were other completely different types of men of a somewhat lower status, who also called themselves surveyors. For example, in London there were paid officials called surveyors who dealt with land and property as employees of the King or municipality (Mills and Oliver, 1967). These were the early ancestors of the borough surveyors who were to come to prominence with the creation of local authorities in the nineteenth century. Also, there were building surveyors and 'measurers' (ancestors of quantity surveyors) who were, at that time, nearer to craftsmen than professionals in status, and belonged to powerful guilds, including those of masons which were strongly masculinist in character (Cockburn, 1977; Knight, 1985). These masonic and technological associations are significant. Although related to a different class subculture from that of the prestigious gentlemen who indulged in estate management, these traditions were to be incorporated within the pantheon of values of the surveying subculture, and thus embedded in the professional institutions that arose from it. They were thus carried through to the twentieth century, and used as exclusionary mechanisms against women.

Enclosure

The fortunes of surveyors improved significantly with the coming of the Enclosure Acts. Thompson (1968: 32) describes enclosure as 'a pre-eminently mappable activity'. The role of land surveying, and especially chain surveying, is a key sensitising concept, and shows a direct subcultural link with 'space' in the model. It is the one bit of essential knowledge that gave surveyors their *raison d'être* to be seen as a profession. To the general public, as stated in Chapter 1, it is still the most visible aspect of their work – but it is not what they do! Chain surveying in particular was retained as an ancient skill that had to be learnt, even when it had long since been delegated to technicians. In due course, the government created its own surveying unit, the Ordnance Survey, effectively removing most land surveying from the private sector, thus making it patently obvious that most private sector surveyors are not chiefly involved in large-scale 'real' surveying.

Thompson comments that even in the twentieth century, land surveying served to remind surveyors of their original links with the landed estates (1968: 191) (a meso to macro link on my model). Over the centuries it gained a symbolic power which could be invoked to repel women on the basis that surveying was obviously 'too technological'. The surveying pole itself became incorporated in the logo of the professional body, the RICS, as a totem representing spatial and professional power. The historical tradition of seeing land as 'sites' to be surveyed, rather than as parts of whole cities and space where people live, may be a contributory factor in accounting for the narrow vision that characterises some surveyors today.

Mapping and plan-making were ways of exerting power over space on behalf of the dominant groups in society who were seeking to formalise their control. The surveyors as the agents of these groups were sharing in this power. Surveying was also a means of formalising colonial conquest and government. For example, George Washington was originally a colonial land surveyor (Thompson, 1968: 44). The surveyor was one of the first 'white men' to be sent out to speed up the process of colonisation. This seemed to give them a natural 'right to rule' which some brought home with them. However, it was not until the twentieth century that town planners, as descendants of the surveyors, sought to use plan-making as a form of power for themselves, as against acting on behalf of the landed classes that their surveyor ancestors had served.

Surveyors found themselves in even greater demand from both landowners and the government following the 1836 Tithe Commutation Act, which imposed new forms of taxation on land, this requiring the drawing up of accurately surveyed maps of every parish. The government saw them as suitably qualified to be potential administrators of the Act. Several of the early surveying firms had been set up by Quakers, such as J.P. Sturge, so they had to think twice before allowing themselves to be involved, as their religion was against both the idea of tithes to the established church, and the taking of oaths as commissioners of the Act. They had to develop a sense of professional distance and impersonality, separating their personal beliefs from their public duties. The elders of the meeting house 'took the common sense view' (Sturge, 1986: 8) that it was perfectly legitimate for a 'Friend' to act as an intermediary between the tithe recipients and tithe payers. This situation illustrates the beginnings of the modern bureaucratic personality (Gerth and Mills, 1954: Merton, 1952) and also reflects a deeper, more worrying characteristic of patriarchal society, namely the division of work and home, so that the rational man leaves his personal opinions at home when he goes to work (Bernard, 1981).

The firm of J.P. Sturge went on to become a major provincial practice retaining their reputation for high integrity to the present day. But the puritanical founder member, John Player, was to gain future notoriety in having his name (used as a trademark) emblazoned across cigarette packets and racing cars in the late twentieth century, as a result of his descendants having married into certain 'capitalistic' tobacco interests in the nineteenth century (a sideways link on the model). As one socially aware woman surveyor commented, 'you can be sure that anything that is seen as a sin nowadays was started by the Quakers in those days', presumably including capitalism, surveying, and smoking in her condemnation?

As government intervention and taxation increased over the years, the numbers of surveyors working for the government, and those seeking to give professional advice on tax minimisation in the private sector, were to grow apace to the benefit of the emerging profession. Indeed, surveyors have often joked that they are always needed under a Labour or Conservative government and they always profit, especially from legislation aimed at reducing the profits from property investment. I could see a parallel between these early surveyors and modern women surveyors. Marginal groups such as Quakers and other non-conformist groups (and later some Jewish families) had apparently gone into surveying as a means of advancement through providing a professional service. They made progress because of their skills and work, rather than because they were initially 'the right type'. However, their role was in a sense that of the outcast, as they were required to assess landowners for the purpose of taxation. Paradoxically, in operating as the agents of the government, with time they became 'insiders' and arguably espoused the values of the dominant group more strongly than its original members (Campbell, 1987).

Although absent officially in these Quaker family firms, women often held a surprisingly equal position in the Friends meeting house (Sturge, 1986: 13). Also, it is rumoured that such women did a considerable amount of what now would be seen as office work, and draughtsmanship at home, and were often their husbands' *confidantes* in professional decision-making. But by the nineteenth century home and work were increasingly separated, as evidenced by a poem written in 1840, on 'The miseries of a land surveyor's wife', by one (*Estates Times*, 13.12.85, No. 825: 9).

Industrial Revolution

Surveyors were in great demand as the spatial enablers of the Industrial Revolution because of the strong territorial emphasis inherent in the development of mines, buildings, canals, railways, and roads. There was a major shortage of surveyors, so much so that young boys were being used to do surveys who could hardly reach on tiptoe to look through the theodolites (Thompson, 1968: 110). Women were told that they could not do surveying because they were too small or weak but this apparently did not apply to these boys (nor as several women have commented, to elderly men nowadays). Unlike in the past when, it is rumoured, some women in surveying families would have done a spot of land surveying themselves, by now most middle-class women were out of the professional labour force (compare Walker, 1989: 92). This is not entirely true, as exceptional women such as Maria Savill, mentioned above, were still doing professional work. It is also rumoured that some daughters in surveying and engineering families were undertaking technical drawing for their fathers.

An added factor, to put in the pantheon of values of the surveying subculture, comes to prominence at this time. The narcissistic love by men of a man-made world full of machinery, engines, railways, and industrialised building structures was reflected in a new macho-technological mentality which infused the value systems of town planning, architecture, civil engineering, and surveying for many years to come, making women feel out of place or inadequate. As stated earlier, Griffiths (1985: 66) comments that technology is even more male than science, and can be used as a major exclusionary mechanism. As the professions became both more technological and less amateur, women's unofficial role was diminished; indeed, the effect of increased professionalisation was to exclude them totally. I have often pondered whether in fact 'professionalisation' is not really a sophisticated form of 'masculinisation', and this cannot be seen as over-reacting in the light of the historical record of the late nineteenth century.

More men were becoming surveyors, and at the same time clearer divisions and status levels within the profession (Weber, 1964: 426) emerged, and the modern range of landed professions was becoming apparent, including architects, town planners, quantity surveyors, land surveyors, and estate managers. However, relatively speaking, these differences were secondary from a feminist perspective since they all formed a closely knit fraternity, as they were all men concerned with the land and property who saw the world in broadly similar terms. Externally, demarcations of territory between the other professions concerned with land and property, such as law and accountancy, had to be resolved. Also there was a growing demand, from both the public and the government, for regulation of standards of professional competence, which eventually led to the introduction of formal entry requirements and examinations. All this occurred between men without reference to women (compare Stewart, 1981). When women were allowed into the professions in the next century they had to fit into these existing male structures. It is another stage again to try to alter professional structures to fit in with the needs of women, and men.

By the mid-nineteenth century, changes in technology and urban theory conspired together to structure the city on the basis of male principles and activity patterns, which were to determine the format of urban development right up to the present day. The new industrial machinery required manufacturing to come out of people's homes, and be concentrated together in the new 'factories' where everyone worked together and did nothing else at the same time (Whitelegg *et al.*, 1982: Part I). With the growth of the railways and, later, mechanised motor transport, many workers sought to live further from their place of work, thus creating distinct residential and industrial districts. The traditional combination of work and home in the same spatial locality, if not under the same roof as in craft industry (and professional practices), had been severed to the disadvantage of women. This separation was compounded when the surveyor's love of mapping was further encouraged by the founding fathers of modern town planning (Cherry, 1981) and transferred into an enthusiasm for 'hygienic' land use zoning, and dealing with social problems by creating orderly spatial layouts.

This trend continued into the twentieth century, town planners further reinforcing their error in categorising urban activities in terms of work, leisure, and home – from a purely male perspective, thus precluding consideration of the fact that women's work was in the home as well as out of it (Le Corbusier, (originally, 1929) 1971: 199; the famous Swiss architect whose work embodied the attitudes of 'rational' European man). Justification for such policies was often based on the assumption that this arrangement increased efficiency (for whom?) and reduced chaos and distractions. Surveyors have never had any reason to object to these principles of town planning. It was in the interests of private sector property owners and investors to have a clearly defined land use pattern, to protect land values from being watered down by less profitable or lower status uses, and to enable infrastructural services to be provided to each site in the most efficient manner. Indeed, many surveyors accept the need for the town planner to act as 'the referee'.

In contrast, to give them their due, many of the philanthropic factory owners who built model communities, whatever their motives, recognised the value to them of women's work, and did in some cases design the layout and the social amenities of their towns to enable women to be workers both in and out of the home (Gardiner, 1923). Ironically, early twentieth century suburban housing was often designed in the style of the idyllic mock tudor cottage that was originally the trademark of such settlements, but without the supporting social structures and planned amenities.

Although such houses, externally, had the environmental benefits of being located in green field sites, they further enforced the principle of housing being at a commutable yet distinct distance from the workplace (McDowell, 1983; Wagner, 1984), this arrangement creating what some urban feminists have variously described as the suburban harem, and the lace ghetto (Hotz, 1977). Internally such property often lacked any space for women. She 'had' either the kitchen or the bedroom, neither of which were really hers. Women were increasingly defined as consumers than producers. From a neo-Marxian perspective, one could argue that both the growth of consumerism and house-building itself were an inevitable result of surplus capital looking for somewhere to reinvest following the decline of the Empire (Bassett and Short, 1980). This theory is of interest as it makes a conceptual link as to how the requirements of the economy actually 'determine' the personal aspirations and values of women and men.

The first wave of feminism and space

The new prosperity derived from commerce and industry and, invested in land and property, brought the services of the surveying profession into greater demand and increased its social standing, whilst at the same time leading to the subjugation of women. Unfortunately, the rising class of the bourgeoisie decided that women should no longer be part of the public realm, or of capitalism itself, suggesting that patriarchy was operating independently of capitalism (indeed this research suggests that *female bourgeoisie* had to wait until the present day, after the *male proletariat* had had their turn, before the women could become the rising class). Women were losing the few property rights they had, in spite of a growth of national prosperity and individual male property ownership. Rich heiresses had to be parted from their money in order to support the investments of their husbands in trade and industry. Hoggett and Pearl (1983: 101) discuss the question of why this was particularly so under English law, when other countries were adopting better systems such as the community of goods within the family, or reserved portions for spouses. In the same way that there is something peculiarly English about having a profession just for land. Married women's property rights (or the lack of them) also reflect this spatial fetishism – the desire for possession of land and property by men.

Not only was a woman's work not acknowledged as part of the economy, or herself as part of society, but women themselves were increasingly seen as a form of property or almost as luxury goods, part of the entourage of the Englishman's home which was also of course 'his castle' (Davidoff and Hall, 1987). This may have originally been an upper-middle-class attitude, but it seemed to have great popularity amongst traditional middle- and working-class men who could bolster their pride as the breadwinner, by stating that their wives didn't have to work and that they were there to cook for them (Campbell, 1985: 111). Occasionally, men surveyors have stated to me, quite out of the blue, 'my wife doesn't have to work' (effecting instant demolition of my self-value, and whatever I had achieved that day) . . .'oh, they just don't think of it that way'.

Increased state intervention by men in the nineteenth century further reinforced the predominance of male values as the basis for dealing with the built environment and creating a 'better' society. The problems of urbanisation and overcrowding were seen as a clear sign that (human) nature was out of control, thus giving rational professional men the right to take control of the lives of lesser beings. This was particularly true of the growing state intervention professions such as town planning, but also true in the private sector in which land and people had to be tamed and made productive by the building of railways and careful estate management. The solution to many of the urban problems of the time was perceived to be through slum clearance and the construction of model housing for the poor, thus emphasising the indispensable role of the landed professions in offering 'salvation by bricks'.

There is a strong class element here as the working class was often seen as the cause of all problems in society. These comments are not merely of historical relevance but are themes that run deep in the psyche of the landed professions even to the present day. There was also a strong, if unstated, gender bias to the theories and ideas that informed men's perception of what was 'wrong', and what needed to be done. Working-class women were often seen as those that had caused 'over-population' in the first place because of low morals and poor hygiene (Richardson, 1876). If middle- and upper-class women were mentioned, they were likely to be seen as either angels on a pedestal or, paradoxically, as neurotic and contributing to the breakdown of society (Durkheim, 1970; Richards, 1977). Indeed, women's selfishness might be cited as the real cause of the evils of society. Seldom does the literature deal with the fact that women might have needs of their own, or that they might see the men as 'the problem'.

In fact, many middle-class women were involved in their own right in reform; however, they have frequently been marginalised and seen as mere 'lady bountifuls'. This was particularly so in respect of women involved in health, housing, and the early town planning movement (Boyd, 1982). Whilst men were praised for their philanthropy and the

The historical perspective

creation of 'gas and water socialism', women were condemned and marginalised for their 'sewers and drains feminism' (Greed, 1987d). This dismissal of women was clearly related to the move for increasing professionalisation and related masculinisation of the government of space in the nineteenth century. However, there was a powerful urban feminist element in existence at the turn of the century both in Britain and North America (Hayden, 1981). Many of the names of key women have been overshadowed by the emphasis on such dominant male figures as Ebenezer Howard, 'the grandfather of British town planning' (Howard, 1960). Then, as now, there were a variety of alternative policies proposed by urban feminists. Broadly, these run along a continuum between those that believe that domestic work should be collectivised and minimised, and those that believe it should remain individualised but that every support should be given to women both in and out of the home. In particular, there were many women interested in housing (Brion and Tinker, 1980) and also in what was called co-operative housekeeping which was meant to cut down domestic labour in the household by applying modern collectivised industrialised methods (Pearce, 1988). Unlike the men, none of these knowledgeable women could call themselves 'professionals'.

The first wave of feminism had a strong spatial and material emphasis (Gilman, 1921, 1979; Banks, 1981: Boyd, 1982) which was reflected in the model communities of the time (Hayden, 1981; Pearce, 1988). 'Material feminism' existed in quite a different form from today, often tied up with utopianism, evangelical reformism, cleanliness, model housing, and less positively, controlling the sinful working classes through the eugenics movement, reflecting a peculiar mixture of class and gender priorities. The roots of many of the values that shape the views of some women in surveying, especially those in housing management, go back to this period (although they themselves may be unaware of this).

Capitalism was not seen as 'wrong' by many such women, but the expectation was that people would use their profits rightly and not selfishly. Such attitudes were often linked to a belief in the rectitude of gaining and using personal wealth in the service of 'mankind', with the successful businessperson operating in a custodial role for others (Ashworth, 1968: 118 *et seq.*). It was considered socially honourable for individuals to have businesses, invest money, and let out privately rented property which was 'safe as houses' (Hinchcliffe, 1988) in order to ensure the well-being of their dependants, and also help 'the poor'. It was considered good for disadvantaged groups to improve their lot through business, 'to pull themselves up by their bootstraps' and liberate themselves. These are all strands that are worth consideration in understanding the antecedents to the social construction of the value system of the modern woman surveyor, 'after all, the state never provided an alternative for women in the public sector, did they?'

Ironically, notable women from this first wave such as Octavia Hill (Hill, 1956) are often disparagingly seen as 'only a housing manager' (which is a statement that requires unpacking in itself). In fact, her ideas were very influential over a wide range of land management issues, including rural planning, and regional economic policy (although she is only begrudgingly given the credit, Cherry, 1981: 53). It is significant that most of the textbooks (which I have to use in teaching town planning) never mention any of the female figures except for Octavia who has become **the** token woman, thus a woman surveyor is deprived of her heritage and also of useful role models. Indeed, some women surveyors and planners I have met who qualified in the 1970s genuinely believed that theirs was the first generation of urban feminists.

The demand for professional standards

In the nineteenth century, anyone (man) could call themselves a surveyor as there were no standards or regulations in existence. The old tradition of depending on reputation as a control on quality no longer worked in the large impersonal cities. Dissatisfied clients were demanding more controls so that they at least got value for money. The government was concerned both with the shoddy workmanship of the building industry as a whole, and with the much wider environmental problems of housing and public health. In 1842 Chadwick had produced his famous report on the sanitary conditions of the new urban working class, roundly denouncing surveyors as inefficient, incompetent, and expensive. He was particularly concerned with the 'weeding out of slow-witted men' (Thompson, 1968: 127). As a result, the 1844 Metropolitan Building Act created a new category of surveyor, the district surveyor, who had to pass an examination before appointment. These are the ancestors of the future local authority surveyors and town planners who were to develop almost separately from the more prestigious, but often less qualified, private sector estate managers and surveyors (Thompson, 1968: 140). Whether they were really benefactors or another 'class' of controllers, as unsympathetic to the needs of ordinary women, and men, as some of the earlier aristocratic controllers of land, is a matter which will be pursued in later chapters.

Incidentally, when dealing with the Industrial Revolution with my students, if I ask what started it all off, the answer will be instinctively that the landowners were the prime movers as they provided the wealth to start the Industrial Revolution and therefore they are to be praised and not criticised. By inference, the surveyors were their advisers and

The historical perspective

responsible for the achievements of the time – rather a different picture from that given by either Chadwick or Marx.

The 1844 Act had not only marked the beginning of regulation but had also established the principle of opening up the profession to people of talent. It was the beginning of making meritocracy (Young, 1958) at least one of the ways in which surveyors might be selected; and in the long run this was to benefit women. At this time there was virtually no surveying education, except for a very few short diploma courses aimed at those entering public service. Most professional education still took place in practice, and degrees for the landed professions were unheard of. Indeed, when education requirements were first discussed in the private sector back in 1834, it was simply suggested (and soon forgotten) that a trainee should serve a four-year apprenticeship 'with his father to the age of 24' (Thompson, 1968: 96)! The move towards formal qualifications was very patchy and half-hearted. Even as late as 1920 it was still a bit hit and miss to know if one had employed the services of a qualified surveyor, as evidenced by the confusing advice to would-be property owners of the time (Ford, 1920: 5).

However, as Thompson comments, 'professionalism was in the air'. The solicitors had combined into a society of gentlemen practitioners as early as 1739. It was not until 1834 that five architects combined to form an institute although back in 1792 a group of quantity surveyors had tried to form together. (Whether estate management surveyors even noticed is another matter.) Nowadays quantity surveyors form nearly a third of the membership of the RICS (Appendix 1) but for many years they had an uneasy relationship with mainstream surveyors, never sure whether they should go in with them or not. Indeed, as with housing, they might be seen as another subuniverse (Berger and Luckman, 1972) but in this case it is based on differences between men, as they have very low numbers of women. Surveyors themselves were not very enthusiastic about forming a professional body as this went against the ethos of the surveying subculture (which, significantly, can exist without any formal organisation). Clutton thought that surveying could only be 'maintained by individual energy' (Thompson, 1968: 132) and projected the role model of the independent professional, which still appeals to many men and women surveyors today. However, they had to be wary of the powers of the more organised lawyers who wanted to take over some of their territory as professional advisers of the landed classes. They also had to strengthen themselves against the potential threat of architects moving in on their territory.

In 1834 nine surveyors set up the Land Surveyors Club, meeting in the Freemasons Tavern, a heavily gendered atmosphere. All had at least some experience of land surveying, but their role as property advisers and valuers was considered more relevant in enforcing their status as professional men (Thompson, 1968: 95) in spite of the misleading name of the club. The aims were to establish their reputation in the eyes of the public, to raise themselves above the level of common surveyors, to work to an agreed scale of charges to avoid unnecessary competition between them, and to develop proficiency by acting as a study group. This early association fizzled out under pressure of work but was recreated in 1868 on a firmer basis. This was set up under the chairmanship of John Clutton with twenty members, with William Sturge being one of the three non-Londoners (Sturge, 1986: 13). Most had fathers and grandfathers who were in property themselves. This core constituted an elite group whose descendants are still highly influential. They in turn selected a further hundred members to join them (Thompson, 1968: 158).

Although they were a voluntary professional association, it was realised that social elevation could only be achieved by better education. Indeed it was hoped that the new Institution would be seen as 'a great university' for the profession of surveyors (Thompson, 1968: 168). Attempts were made to introduce a programme of learned papers, but as Thompson remarks, (1968: 170) these were more 'suited to a ladies' literary society than a professional institution'. However, they were aware that it might be prudent to be seen to have examinations. In 1881 the Institution was granted its Royal Charter on condition that examinations be introduced. These ran virtually unchanged until 1913 when the examinations were further subdivided and modified, but basically the 1880 structure was adhered to until 1932! The surveyors were rather proud of the fact that they were the first 'rule of thumb' profession to introduce examinations (although they were not compulsory on all members, especially not the founder members). So by 1881, the basic format, chartered status, and examination system had been established to take the surveyors into the twentieth century. Sturge produced a seminal paper on the education of the surveyor in 1882 (Thompson, 1968: 182) which actually stated that it was impossible to test the professional knowledge of a surveyor as his work was of necessity practical rather than theoretical. This 'healthy' anti-education attitude has continued within the subculture up to the present day. However, there seemed to be a new enthusiasm in society itself for education, albeit for the working classes, as evidenced by the 1870 Education Act. By the time of the 1902 Education Act, it had become more acceptable to have standards for middle-class, and eventually for professional, education too. Indeed education became part of the spirit of the age. This seemed rather late, as after all the entire Industrial Revolution had occurred without more than the minimum of formal education.

Chapter four

Twentieth century development of surveying

Introduction

At the turn of the century, the Surveyors Institution had only 2,267 members (Thompson, 1968: 341). A few surveying courses were beginning to develop. It is significant that certain agricultural colleges were among the first, thus reinforcing the traditional rural (and arguably 'male') associations of the subject. One such college was well-known for not admitting women until the early seventies. One woman told me that when she enquired about entry in the early 1970s, she was told that they would like to take her but there was no accommodation available for women, although admittedly the situation has much improved in recent years. The first full-time university degree in surveying was introduced at Cambridge in 1919. Surveyors were concerned about status and saw 'education' as rather a down-market activity with working-class associations, unless it was Oxbridge. They were worried that young surveyors would acquire 'desultory and expensive habits, if not vices' from university life which would spoil them for the drudgery of the office. However, it was suggested that surveyors should be educated to hold their own with clients (Thompson, 1968: 229). (Compare with the idea that wives should be educated to converse with their husbands, Bernstein, 1975.)

Private correspondence and crammer courses were the main methods of qualification. It is unknown whether any women attempted these courses. Several women who qualified in the last ten years through the now diminished correspondence route, told me they thought it best always to use initials rather than first names on essays. 'Postal domination' was to characterise surveying education for many years with college courses being the exception rather than the norm. In the interwar years the old pupillage system began to crumble, which in the long run was to the advantage of women in providing more open access, making the possibility of education and qualifying experience no longer linked to the whims of the employer.

The early Surveyors Institution had established entrance requirements (at approximately GCSE level standard) which entitled a candidate to be articled to a practice from the age of sixteen. The range of subjects required for entrance was only available at public and good private schools in the late nineteenth century, and so one needed 'class' as well as 'brains'. The social connections and sports activities participated in at school have always been seen as equally if not more important than academic pursuits. All male surveyors are not equal; a ready-made status hierarchy based on school background and sport exists between men in surveying even before they enter the profession. Women's (lower) position must be seen against this already existing pecking order.

Whatever the educational backgrounds of surveyors, many have always liked to make much of the fact that they are highly educated professional gentlemen (Steel, 1960). An impression of strong university links is given, either with the technological side of surveying science (Hart, 1948) or with the classics (Battersby, 1970[1]). These themes continue through the present day, being reiterated in the 'Presidential address' each year and in other articles too numerous to mention. Surveyors often use 'borrowed glory' by publishing articles by distinguished outsiders, especially from the legal profession (Heap, 1973). Heap (1973) projects a sense of humility and suffering in implying that a professional is only 'a man who does his duty'. Incidentally this contradictory image, that 'it's nothing really, it's quite ordinary to be a surveyor', is a theme which often appears when 'society' or 'women' threaten or question the professions.

The entrance of women

There were no urban surveying courses at the turn of the century although the City of London College, and Birkbeck College from 1919, each offered a few classes. Thompson describes (significantly within his chapter, 'The Institution and the Public') how in 1899 a Miss Beatrice Stapleton asked to take the exams from Birkbeck College, presumably having attended the course (Thompson, 1968: 318). With relief the Institution stated that candidates had to be employed in surveying first for a specified period, and that it was impossible for a woman to find such employment. It did the same in 1915 when a Miss M.V. Smith applied, and declared that women could not become surveyors; nowadays one is more likely to be told that one cannot be a woman and a surveyor at the same time. However it did concede that its charter merely spoke of 'persons who are surveyors' and did not specify their sex.

Nevertheless, some women appear to have been taking surveying courses, as an issue of *Every-Woman's Encyclopaedia* (1911: 2840) mysteriously contains a photograph of women attending a surveying class at University College, Reading, about 1910 (there was no surveying course there then). It is captioned, 'Students attend a lecture on surveying, they are constantly taught the doctrine of hard work, and are prepared early for future arduous conditions to which their professions will subject them'. It is a classic case of women mistakenly perceiving the level of standards required and therefore imposing on themselves the need to work twice as hard (a false image which some men were/are only too happy to foster).

In 1918 an external degree of London University in estate management was established and taught mainly through the newly established College of Estate Management. The link with the University was made because William Wells, a leading surveyor of the time, was the brother of the Vice Chancellor. In 1968 the College moved to Reading University, and it is interesting in respect of 'surveying dynasties' that another Wells was President of the RICS at the time. At a more mundane level there was a gradual growth of courses in technical colleges, especially the pre-war first wave polytechnics which operated on a night-school basis (Venables, 1955: 293) and provided the bulk of surveying education. Education itself did not operate as an exclusionary mechanism or as a conveyor of status; rather, the requirement of having sponsorship and experience from an approved office was much more of a hurdle through which closure could operate.

Meanwhile, the School of Civic Design was set up at Liverpool in 1909 and in 1911 a Chair of Town Planning was endowed at Birmingham University by Cadbury, the famous town planning philanthropist and erstwhile chocolate manufacturer (Ashworth, 1968: 193). The Town Planning Institute (TPI, now the Royal Town Planning Institute) had been established in 1913 (Thompson, 1968: 193) initially as a professional rather than qualifying association. Town planning, with its somewhat Utopian reformist ideals, appeared to represent another subculture from surveying, but in fact many notable surveyors supported its creation and joined the new body. However, surveyors have always looked upon town planning with some misgivings seeing it as 'un-British' and rather radical (Rowe, 1955) and possibly linked to garden cities, bicycles, liberated women, and Fabianism; although there were very few women officially visible in the TPI. As will be seen, since its inception, it seems some surveyors have either tried to be in control of town planning, or to dismiss it. It was an affront to their claim to be the main landed profession. Town planning was obviously urban and progressive, whilst surveyors were still tied to the past.

Although in the nineteenth century women had been pushed out of male employment areas, after the First World War, because of economic necessity and first wave feminist pressure, women were allowed into the professions, following the 1919 Sex Disqualification (Removal) Act (Thompson, 1968: 319). It was believed that women would never qualify as surveyors because so few offices would take them. Undeterred, Irene Barclay and Evelyn Perry, who were to be the first two women surveyors, attended an evening course in surveying (Barclay, 1976: 15). I was able to correspond with Irene Barclay, who was not only the first, but until her recent death the oldest, woman surveyor (27.5.1894–21.3.1989). She remembers that they were the only 'girls' amongst a crowd of young men, and that the lecturer who dealt with drainage and sanitation was acutely embarrassed (Barclay, 1980; Obituary, The *Guardian*, 1.4.89). (I myself had the misfortune to come top in the sewerage and drainage examination at college, and was dubbed Queen of the Sewers ever after, and so I understand fully how odd men can be about these matters.) Irene was first to qualify in 1922, rapidly followed by Evelyn. She got around the practical experience aspect by being employed to work on working-class housing estates by the Crown Estates Office. She also appeared to be supported by a network of professional people, including sympathetic men surveyors within the Institution.

Soon after qualifying she set up on her own as a chartered surveyor, got married, and had children at the same time. She did all the things that many women surveyors do today and experienced all the same problems. No doubt some of the men felt threatened by the entrance of such dynamic women. Some men almost seemed to fall into surveying, as it was seen as ' jolly good thing to do' (*Chartered Surveyor*, August 1978, Vol. 111, No. 1: 8). (This is less true today, and of course there have always been 'bright' men too.) Women meanwhile had to be very hard-working and bright, and still experienced great difficulty. Women's work seemed to be judged and perceived quite differently from that of men. No wonder women often feel there are two sets of rules and two separate surveying cultures, one for women and one for men.

Irene Barclay was an example of the early type of woman surveyor who went into it because she was from a family deeply committed to social service with a liberal non-conformist background. In one sense she is what might disparagingly be seen as a 'lady bountiful' type who was now permitted to work from within the profession, who would have previously done similar work outside the profession. She is also a 'new woman' or even a bourgeois feminist, who wanted her own career. Interestingly she is often seen as 'only a housing manager' although she undertook the full range of professional work (Power, 1987: 28). On her retirement 50 years later she stated that she regretted that more women did not follow her example by going into surveying (*Chartered Surveyor*, January 1973, Vol. 105, No. 7: 342) although she was glad that many became housing managers. She clearly 'could see it all' but ironically many modern women surveyors, who would benefit from her wisdom and example, barely know of her, as she had become 'invisible' in her own lifetime.

The surveyors responded rapidly to this influx of women; there must have been at least twenty on their way to qualifying by the late 1920s, and a male membership of 5,305 in 1926 (Thompson, 1968: 341). In 1931 the Institution established a special certificate for women housing managers (Power, 1987: 29), an occupation which, according to the surveyors, had only developed as a distinct profession since 1919 (Thompson, 1968: 318). Well, what a coincidence! This is a classic example of diverting women away from the main pitch by creating a special niche for them, a tactic which is still being played out within the tangled web of the profession to determine the gender demarcations within surveying. Housing managers were not counted as full members of the RICS (and their names did not appear in the Year Book). Of course it is not true to say that the housing profession had not existed prior to 1919, nor is it true to infer that it is somehow softer and more feminine, as similar technical and estate management skills are required as for 'ordinary' surveying.

A minute core of women was now established inside the landed professions and this may have given women a marginally greater advantage than they had outside the professions. It should not be forgotten that in the 1920s, the 'new woman' might be actively involved in social reform and political activity without necessarily thinking in terms of personal success or professional status (Perry, 1987: xii). However, professional women were often considered twice as eligible, because of their qualifications, to resume what was seen as the traditional and acceptable role for educated ladies, to sit on various committees on matters related to social reform. Irene was an active member of the wartime study group on the condition of 'slum-dwellers' made between 1939–42 (WGPW, 1943). Her efforts and much of the data were totally forgotten. If men had done it, it would have been seen as one of the great social surveys of all time.

A digression into housing management

Housing management contains a high proportion of women and one can observe certain mechanisms at work within this specialism which are less visible within more male-dominated areas of surveying. It may be argued that, far from women only recently entering the landed professions as a result of being given greater rights and encouragement by men, in the case of housing women have actually lost what was previously a strong presence as a result of being pushed 'down and out' (compare Power, 1987). This is not a contradiction, as men can encourage more women to enter a profession and recruit more female students, whilst at the same time others are blocking the progress of women already in that profession, thus creating a clear vertical gender division. This is why, in research of this nature, one has to be very careful to consider the qualitative implications of apparent quantitative improvements.

Housing is only a small, but growing, specialism within the RICS. Housing managers may also qualify through the Institute of Housing which is separate from the RICS, and a much smaller and arguably less prestigious body (Appendix 2). I am indebted to the help of the Institute of Housing, and Mary Smith, one of its few women past presidents, for elements of the following account, although it should not necessarily be seen as reflecting their views as it incorporates my thoughts too (Smith, 1989). Octavia Hill is generally accepted as the founder of modern housing management (Hill, 1956). In 1865 she took over some run-down houses in Marylebone and improved and subsequently managed them. She worked on the principle that people and their homes had to be dealt with together and not separated. She believed that both landlord and tenant should respect each other's rights and have strong moral duties to each other (a far cry from the 'tenant right' and 'landlord wrong' conflict of later years). A weekly collection of rents by trained women housing managers gave them the chance to get to know their tenants and their problems. Many older women have told me that as men took over housing management they lost this personal touch, and tended to deal with the impersonals of buildings and budgets to the detriment of the tenants.

In 1916 the Association of Women Housing Workers was founded, later renamed the Association of Women House Property Managers. In 1927 the first posts for women housing managers in local government were established, but already men were becoming more interested in the area. State intervention in housing was changing the nature of housing management and making it a major bureaucratic (male) function which, of course, men entered at a 'higher' level. In 1932 the Society of Women Housing Estate Managers was formed, the word 'estate' being dropped in 1937. Also in 1932 the RICS introduced the Housing Managers Certificate to formalise their housing qualification. In 1948 it became known as the Society of Housing Managers, at which time men were also admitted although its membership was still mainly women.

In 1931 local government officers (mainly male), who were often without any formal training, created their own separate Institute of Housing, and made no attempt to join the women's organisation. They administered the increasing levels of state intervention in housing (as against charitable private initiatives), allegedly on behalf of the working

The historical perspective

class, which was chiefly defined as men by other men. For example, the post-war 1919 Housing Act had specifically aimed at providing 'Homes for *Heroes*' (Swenarton, 1981). The Healing Committee Report (1939) on surveying education (the first of the big reports to be produced by the RICS on the subject) stressed the need for special housing manager surveyors and construed these in a male form, as if the previous history and experience of women housing managers had scarcely existed. In 1965 the Institute of Housing (male) and the Society of Housing Managers (female) amalgamated to form the Institute of Housing Managers. In 1974 this became the plain Institute of Housing again. Many women, whom I have spoken with, believe that an unlikely combination of the 'male left' and the existing bureaucracy-oriented 'local government fraternity' had effectively recreated 'housing' in their own image as if it had been a 'new' subject, apparently ignoring many existing women and their ideas whilst pretending that they were in favour of opening the profession up to women. In reality, many of the women saw themselves as far more radical than such 'socialist' men, who were perceived as really rather 'conservative' and 'narrow'. In the late 1970s, housing was given a much higher profile within the RICS itself, when after years of a 'special' membership category, it eventually became a full option within the mainstream general practice division. Several surveying courses now give joint RICS/Institute of Housing exemptions, and some believe the RICS will eventually 'gobble up' the Institute altogether, with interesting implications for women!

Control over the built environment

State intervention gave men the power, that private practice never had, to impose their ideas on a massive scale on behalf of 'the people'. The twentieth century had brought a new enthusiasm throughout Europe to create a future based on technology and science, and to sweep away the apparent clutter of the past (Pevsner, 1970). This led to a wave of new architectural movements such as futurism and functionalism which manifested themselves in the design of high-rise buildings, especially multi-storey blocks of apartments for the working classes. Individualism and private ownership were seen as regressive, and houses were to comprise factory-produced units built for maximum efficiency. For example, Le Corbusier is infamous for his statement, 'a house is a machine for living in', which, as many women have commented, could have only been said by an elderly bachelor who had no idea how ordinary people lived (Ravetz, 1980).

Such architectural opinions were often inspired by a belief in the importance of creating an 'efficient' mass production economy, which itself was based on male definitions of work and production. In contrast, consumption (arguably woman's realm which included activities such as shopping and home-making) was often seen as trivial, bourgeois, and self-indulgent, dismissed as an evil side-effect of 'capitalism' and not seen as an essential activity to sustain life and service the needs of the workforce. It is important to point out the existence of such 'out of touch' attitudes, as they were to colour the world view of generations of twentieth-century urban managers including town planners, housing managers, and even some surveyors too in the public sector, guaranteeing that little importance would be given to women's needs. On the other hand, commercial surveyors who acted as advisors of private sector developers had much more respect for consumption and the business world itself, but this was not necessarily to the benefit of women, but it often was seen as such.

The inter-war period was one of great urban growth, including the spread of semi-detached suburbia. Quite apart from 'real' surveyors, there was great opportunity for estate agency and a whole proliferation of quasi-professional chaps in 'property' who might have felt no need to go on and become full professionals. There was still plenty of spare capacity in the market and no need for exclusionary mechanisms to restrict entry to the profession.

Many would say that the small builder, or the much maligned speculative developer of the suburbs, actually went in for more direct consultation with his future residents, because he was local and he had to please his clients or they wouldn't buy his property. Significantly, many such builders actually lived in the same type of house in the same street as their buyers, which is something the modern architect seldom does. Also, the local estate agent was more in touch with community needs than the distant higher status surveyor. Increased formalisation of the education of surveyors and growing professionalisation isolated them from the community they served. These are issues of status and demarcation of work between men which nevertheless had disastrous effects for women as consumers of the built environment. Relatively speaking, the track record on housing by the state and its experts in disregarding the needs of women might be one reason why many women prefer the private sector (although private and public sector are arguably two sides of the same patriarchal coin).

Post-war reconstruction

Following the Second World War, the new Labour government introduced a programme of post-war reconstruction, in which construction was the operative word. Emphasis was put on planning of both the economy and built environment, supposedly, on behalf of the working class. One would have imagined that the private-sector-oriented

The historical perspective

conservative surveyors would have been up in arms. Not a bit of it! As members of the main landed profession, they threw themselves into the task of implementing the policies of what they no doubt saw as an unacceptably socialist government with complete dedication and professional neutrality. They did quite well out of it too, either as government employees, or as the advisers of private clients seeking to evade the effects of post-war reconstruction on their land and property.

Most of the post-war planners were originally surveyors, engineers, or architects (Marcus, 1971). Many more surveyors than today were working in the public sector, in valuation, estate management, town planning, and other specialisms. Even in the late 1960s, over 40 per cent of surveyors were to be found there (Thompson, 1968: 350) compared with less than 20 per cent today. They seemed genuinely enthusiastic about many of the more 'socialist' aspects of town planning. After all, since the Labour government had taken much of the power away from the owners of land, this gave the landed professionals themselves greater power to control space (in their new incarnation as 'town planners') in the way they wanted without the hindrance of a private commercial client, or the lord of the manor telling them what he wanted (compare Orchard-Lisle, 1985[2]).

The role of the surveyor in town planning continued to be a major issue throughout the post-war period as evidenced, for example, by a policy statement on this (RICS, 1966) in which it is implied that surveyors should have a prominent role. Indeed, one might argue that the nature of post-war planning and urban development owes more to the influence of surveyors than planners (compare Burke, 1980), including the more controversial aspects such as town centre redevelopment, and shopping provision (Shepherd, 1954; Holford, 1949) and slum clearance (Macey, 1958; Trepas, 1970). Phrases such as 'the speedy removal of the residents from the area will facilitate its efficient renewal' were commonplace.

However well-intentioned these surveyor planners were, they had very little idea how the ordinary woman or man lived, and frankly they felt much happier designing spatial solutions (Keeble, 1956) than worrying themselves with the complexities of human beings. In fairness, however, many surveyors were at their most enlightened and liberal in the post-war years. I was impressed with articles in the immediate post-war surveyors' journals that, in quite a matter of fact way, assumed that nurseries for working women would be an integral part of new residential area plans (Chapman, 1948: 219). None of the asocial commercialism of the present day is in evidence in the immediate post-war period (Lane, 1958).

The fatal mix of a desire to plan for society on the basis of scientific rational methods combined with badly informed men with power to make decisions for others was to disadvantage women again and again (Greed, 1987a). For example, in the new towns, planners, using the neighbourhood unit concept 'for the community', often put shops at the opposite end of the estate from the schools and then zoned the factories on the outskirts. All this involved a considerable amount of additional walking for many women, as they rushed to and from work to collect their children from schools (whose hours were also seen as most inconvenient and impractical by many working families). There seemed to be a hypocritical conflict of attitudes towards women. The housing design was based on a watered-down version of middle-class suburban estates where women were not expected to work, but at the same time industrialists were often attracted to the new towns by the promise of cheap flexible part-time female labour. Such issues are nowadays frequently discussed by urban feminists, as in a paper by Judy Attfield entitled 'Inside Pram Town' (1989) in which images of 'mud and babies everywhere' vividly describe the environment. As had been evidenced in the war effort, men did know how to plan more sympathetically for the needs of working women, but they chose no longer to do so (Wekerle *et al.*, 1980).

There have been a few attempts to study retrospectively some of the most well-known plans of the twentieth century to find out what they have to say about women, which frankly is a matter of looking for what is not there (Morris, 1986; Roberts, 1988). Both general feminist commentators (Wilson, 1980) and urban feminists (WGSG, 1984: 58) regret that a great opportunity was lost in not accommodating the needs of women, as well as men, in post-war reconstruction.

Town planning problems were defined in terms of land, not people, in this period. If the planners and geographers saw suburbanisation as a problem it was not from the perspective of inconvenience to women, but because it constituted what they saw as urban sprawl (Hall, 1977). One of the main policies of planning right up until very recently was to prevent this sprawl (the present government has astonished everyone by declaring (Circular 16/87, *Development involving agricultural land*, Department of the Environment) that there is now a surplus of agricultural land owing to overproduction!). Indeed, town planning law, by its very nature, is much better with dealing with problems when they are presented as matters related to land rather than people (Morgan and Nott, 1988: 139). Whether or not one should see this as part of some intentional conspiracy or the result of the spatial background of the planners who conceived the post-war system, it has meant that many women town planners and other interest groups have had great difficulty packaging their people-related policies into the required spatial format which will stand the test of legal appeal (Taylor, 1988).

Women had been allowed into several areas of public service sur-

The historical perspective

veying during the war; indeed, many unqualified women had been effectively carrying out professional work, but now of course the men wanted 'their' jobs back. Some areas had remained male territory and were closely guarded until such time as the men had reason to change their mind. For example, for the Ordnance Survey, it was commented in 1950 that 'women should be eligible to sit for these examinations, although in view of physical and other limitations there will be a limit to the number who can be so employed at any one time' (*Chartered Surveyor*, October 1950, Vol. XXX, Part IV: 288) – an illogical muddle of time limited by numbers. Nowadays in contrast, women are increasingly welcomed into 'real' surveying. I spent a day on site with a woman land surveyor triangulating, and thought how quietly and unassumingly she was doing this highly skilled work, the very activity which originally had been the basis of surveyors' claims to professional monopoly. But for her it brought no great status (although she found the work very interesting in itself); indeed, nowadays one 'only' needs a non-degree qualification to do many aspects of this work. To unpack this, the skill has not diminished – only its social construction and its gender and class associations have changed. Clearly, here, culture rather than biology is at work. In fact, as rumoured to me by other indignant men, some men were actually entering surveying at this time after the war in order to escape the opportunity to be real men – draft-dodging in fact (*Chartered Surveyor*, November 1948, Vol. XXVIII, Part V: 247).

The 1960s

Planning in the 1960s was based on assumptions of economic growth, increased affluence, and rising car ownership. Needless to say that many women still do not have cars. Planning cities for cars caused great inconvenience for pedestrians and users of public transport, thus adding fuel to the fire of the emerging urban feminist movement as women began to see themselves as a group that was discriminated against by the car-borne white middle-class male planner. Indeed, there was growing dissatisfaction with the policies of the planners from a variety of directions. In particular, many people were organising community groups in protest against their houses being demolished to make way for urban motorways and car-parks to get the increasing number of suburban commuters to their offices. Planning and the private sector went hand in hand. There was a gradual upturn of the commercial market in the 1950s rising to the crescendo of the property boom (Marriott, 1989) of the late '60s and early '70s. In spite of the vast increase of town planning powers, the private sector was able to carry out many questionable town centre redevelopment schemes with the full blessing of the planners and indeed sometimes in partnership with them. The surveying profession itself expanded in response to this property boom. Paradoxically, poverty and unemployment were re-emerging, especially in what came to be known as the inner city (some knew that they never went away).

This was reflected in academic literature in a move away from a consensus view of society to greater emphasis being put on the existence of urban conflict and on an acknowledgement that everyone's needs were not the same. The planners and urban managers themselves were studied as 'the problem' rather than assuming certain deviant groups in society were the primary cause of trouble as in past urban studies (compare Pahl, 1977a, with Morris, 1958). Later, more radical studies of the urban process and the professionals involved emerged (Harvey, 1975; Bailey, 1975, and in due course, Dunleavy, 1980; Goldsmith, 1980; Simmie, 1981) which showed that all was not fair or value free (even without considering gender).

The 1971 Town and Country Planning Act introduced a new type of development plan which was based on presenting high level goals in the form of an abstract policy statement, illustrated by diagrammatic non-site-specific plans which were produced on a computer grid format. Paradoxically, the same Act required a greater level of public participation as 'planning is for people' (whether that included women is open to debate) (Broady, 1968; Skeffington, 1969).

Development of surveying education

The growth in surveying education and the profession itself is a reflection, if not a result, of these changes in land use and development (another link between space and the surveying column in the model). In 1930 the Surveyors Institution changed its name to the Chartered Surveyors Institution (CSI) and the term 'chartered surveyor' was more commonly used, reflecting a rise in status. It remained the CSI until 1946 when it took on its present name of RICS after the prefix 'Royal' was granted (Thompson, 1968: 333) at which time the membership totalled 7,805 (Thompson, 1968: 341). Women's true position in the inter-war years may be gauged by such gems as the fact that a major discussion took place among surveyors on 'whether the presence of ladies was consonant with the dignity of professional gatherings and social events' (Thompson, 1968: 255). One cannot generalise as, for example, the first woman quantity surveyor, Margaret Mitchell, was entrusted with doing the Bill of Quantities for the Cenotaph War Memorial in London (*Chartered Surveyor*, November 1968, Vol. 101, No. 5: 250), but I could find the names of less than fifteen women listed in the 1939 'Register' (the Year Book, RICS, 1939), among a total membership of around 7,000 (Thompson, 1968: 341). As regards salaries, both women and men started on the same rate in public service, but men's increments went up much higher (Willis, 1946: 47 *et seq.*), and no doubt it was presumed that women would leave. Men, then as now, got additional increments and pension rights in respect of war service, but women received no pay for years taken in creating life.

Surveying education was becoming less amateur. In 1913 the final (and only) examination had been divided into two to create an intermediate and final level. In 1932 the intermediate level split again, thus making three levels. This remained unchanged until 1966 when 'A' level entrance at 18 years of age was established. The Healing Committee (1939) improved and rationalised the examinations into three equally spaced stages, and also stated that candidates must have four years of experience and be student members of the Institution to be eligible for examination (Thompson, 1968: 189). In 1939 there were 300 student members (compared with around 7,000 full members) but, as a result of this new ruling, student numbers rapidly increased to 2,000 in the war years, rising to 5,000 in the 1950s (compared with 10,079 full members in 1951, Thompson, 1968: 341).

Education was a key theme of post-war reconstruction as evidenced in the 1944 Education Act. There was a concern for efficiency and the raising of standards in professional education (Langdon, 1949). The Schuster Report on Qualifications for Planners (1950) was to provide guidelines for surveying education, although its original intention was to set up a new range of separate town planning courses as attempts to implement the post-war legislation had showed up a massive shortage of educated 'man'power. Perhaps the Labour government had realised too late that it was foolish to use conservative-oriented surveyors to carry out Labour land use policy; a new landed profession of more politically sympathetic planners had to be created. Of course, at the same time, many keen women who had produced plans in the war were being sent back home, or to the drawing-office to resume their previous life as 'tracers' (as women draughtspersons used to be called). As Colin Ward (1987) comments, 'she was resented because she knew too much about other people's jobs'.

Although surveyors became much more concerned about education in the post-war years, they still wanted to have it both ways. The Watson Report (1950) on the 'Educational Policy of the RICS' said that full-time education should be encouraged but that the majority would continue to qualify part-time. It was simply not part of the subculture of surveying to be a full-time student; the avoidance of, not the participation in, education confirmed subcultural values. The Wells Report (1960) welcomed more students from higher education into surveying, but still seemed to assume that the main route would be part-time and non-university. It includes the statement 'the occasion will demand men and **women** who are not only able practical surveyors, but also capable of original imaginative thinking'. This is the first report to acknowledge the existence of women surveyors. It also introduced the idea of 'CPD' (continuing professional development), a subject dear to the heart of the modern surveyor. Not only should surveyors undertake an approved course of study but they should go on learning in practice, continuously updating their practical knowledge. This may be interpreted as an additional modern exclusionary mechanism that women might particularly encounter as a problem; or as a form of educational inflation or, more benevolently, as the last trace of surveyors trying to put their traditional emphasis on practical experience back into a world that prized formal education above practical experience.

Student members reached over 7,000 in the 1960s, most of whom were taking the part-time or 'personal' study routes to qualification. This represented well over a 2:1 ratio of members to students (for example, there were 15,557 full members in 1961 and 18,290 in 1966, Thompson, 1968: 341), reflecting the growth required to 'man' the postwar property boom. However, numbers of part-time students gradually declined relative to full-time courses, although the overall numbers of students in both sectors increased. Numbers of students doubled between 1969–79. In 1979 there were 14,400 students, compared with 40,818 full members whose numbers had also doubled in the same period (SITE (Surveying In The Eighties) Report, 1980).

Passing the examinations does not appear to have been as important as being seen to be taking them whilst in employment as a trainee surveyor. The Brett-Jones Report of 1978 states that there was on average a 40 per cent pass rate between 1945–77, and before the war the figure was even lower. Credentialisation of itself was not the key to professional power. Right up until the 1970s, education was still only seen as an aid to passing examinations and not as a valuable experience in its own right, and many did their best to put off the evil day when they formally qualified. Education was certainly not a major factor in either conferring or confirming subcultural values or in career advancement. Indeed, the Institution itself seemed to be of limited importance in this respect, being a loose confederation with only a monthly journal in those days. The professional office was still the centre of the subculture, therefore it is hardly surprising that even nowadays women find that although apparent equality has occurred in education, that does not matter relatively speaking because the true professional socialisation process occurs in the surveyor's own domain within the office.

In 1960 only 10 per cent of applicants had 'A' levels, and 75 per cent were qualifying by correspondence. The situation was not reversed until 1983 when two-thirds of surveyors were qualifying by full-time courses. But this is not to infer that surveyors were any 'worse' than other professional groups, as in the past many others such as accountants, lawyers, architects, and town planners had similar entry requirements and were articled straight from school. What mattered was whether one had the right contacts to get into the right office. Indeed, it was not uncommon for people to pay to get in and to receive no salary for several years. College education did not have the same importance as, for example, in North America; to be 'academic' might even be a disqualification. Perhaps women (and working-class meritocratic men) were playing the wrong ball game in projecting their own assumptions about the importance of qualifications and grades onto the question of eligibility for the professions. Whilst men surveyors wanted less education, women in society wanted more. Obviously the power did not lie in education.

Then from about 1967 everything changes. A whole issue of the *Chartered Surveyor* is devoted to education in June 1967 (Vol. 99, No. 12) to be followed by several equally important articles in 1968–9 (for example, French, 1969). The 1960s was the age of the white heat of technology with a Labour government intent on creating an efficient education system to match (Robbins, 1963). Dull old feudal surveying 'chameleon-like' as ever (Teale, 1985: 4), now presented itself as a worthy case for special treatment, as a 'technological' profession, in spite of having long since lost many of its associations with surveying science. The one exception to this was quantity surveying, which had grown out of all recognition in the 1960s as a result of the property boom of the time with its emphasis on high-rise construction for which their expertise was particularly relevant. The new white heat of technology image was as much a barrier to women as the old landed gentlemen's club image.

In addition to Robbins, there was a series of secondary reports that had a bearing on surveying education including the Hennicker-Heaton Report (1964) on day release, and the Pilkington Report (1966) on technical college resources, culminating in the 1966 White Paper proposing the creation of the 'new' polytechnics (Ministry of Education, 1966). The RICS itself produced a series of reports including the Eve Report (1967) which suggested that in the future full-time education would be the normal route to qualification. However, it also laid the foundation for the 'TPC' (test of professional competence) which all surveyors are nowadays required to take after a period in practice before they can become fully qualified. Following on from this, the Percy Report (1970) on higher technological education defined the future parameters for surveying education, and in 1972 the new polytechnics were created (Robinson, 1968) as the institutions where the expansion of professional education was to take place. Town planning education was also being revamped from a somewhat different ideological perspective. The separation of planning and surveying was to have major implications regarding surveyors' influence on 'what is built'. Some surveyors nowadays seem to assume that the deficiencies in surveying regarding lack of social concern should be picked up and rectified by planners at a later stage in the development process. This would free surveyors from any need to worry about the social implications of their work so they can concentrate on getting the best financial return for the client. In a sense, planning was to become the compensatory profession doing the 'emotional housework' that surveying was too busy to deal with – a female role to surveying's male image? (This is notwithstanding the fact that about 97 per cent of planners were male in those days, and today the figure still hovers around 85 per cent.)

I am indebted to the work of a student from the early 1970s who chose the topic of surveying education as his final year dissertation topic (Davies, 1972). Among other matters, Davies even investigated the numbers of women students on surveying courses at that time, which he found to be about 1–3 per cent depending on the course. I well remember having it impressed on me as a student elsewhere, at about the same time, that there was no problem and I was just being silly in wondering about how many other women there were.

Only about 1 per cent of surveyors had cognate (i.e. degrees in surveying) degrees in 1928, and it is estimated 2 per cent had degrees in other subjects. Most of the rest were doing College of Estate Management courses, local evening classes, or studying on their own whilst they worked. Even by 1955 only 10 per cent had cognate degrees rising to 12 per cent in 1963, and then rapidly to 67 per cent in 1985 and now reaching over 75 per cent (Wareing, 1986) as a result of the changes in higher education that occurred in the 1960s (*Chartered Surveyor*, 14.7.83, Vol. 4, No. 2: 101). Full-time surveying students comprised fifty in 1945, 304 in 1950, and 450 in 1967 (Thompson, 1968: 221) but there were also 3,744 attending part-time courses in the late 1960s (Davies, 1972: 26) plus many doing correspondence courses, or 'nothing much'. In 1966 the RICS approved seven new degree courses and nine diplomas (a degree nomenclature was less common then) in general practice surveying, in nine different colleges; the number of colleges where new courses were run rising to twelve in 1968, and to fourteen in 1972. Including pre-1960s' courses, by 1972 there were in total nineteen colleges offering a range of full-time degrees and diplomas in all aspects of surveying, and many more offering part-time routes (Davies, 1972; *Chartered Surveyor*, May 1969, Vol. 101, No. 11: 550). Several of these courses were validated as CNAA degrees (Lane, 1975) being located in the first thirteen new polytechnics which were created in 1972.

There were now around thirty-six colleges in Great Britain (*Chartered Surveyor*, 1.9.83, Vol. 4, No. 9: 418; RICS, 1987a) offering, between them, approximately ninety full-time courses (including sandwich ones) counting all the surveying specialisms, with over twenty-four of these courses being general practice and estate management courses. Of the thirty-six colleges, only ten are universities, twenty-four are polytechnics (or equivalent), and two are military establishments (80 per cent of surveying courses are in polytechnics). In 1985 there were 1,550 students in universities, 6,400 in polytechnics, and 1,400 part-time students mainly in technical colleges and some polytechnics. There were also 4,000 direct students, mainly correspondence with the College of Estate Management including many retake students from a wide range of age groups (Wareing, 1986). Significantly this route continues to flourish as the College of Estate Management registered 5,344 students in 1989, a 15 per cent increase on 1988 (*Estates Times*, 5.5.89, No. 993: 7).

The establishment of this new and vastly expanded system was to set the framework for the future qualitative development of surveying education. It was not the intention to specifically benefit women in all this; indeed the purpose of the polytechnics was originally to benefit working-class boys (Whitburn, 1976). In reality, the two groups that benefited were middle-class women, and upper-middle-class males. Many of the principles established by the CNAA and the new polytechnic courses were to be echoed in the Brett-Jones Report, 'Review of Educational Policy' (1978). The following interesting statement is found in paragraph 2.5 of this document, 'perhaps the most difficult problem is the conflict between the profession's need to provide for its future manpower requirements at partner and principal level as against the employer's need for immediate technical assistance in his office' (was this a prophecy regarding the future male/female division in the profession?). There is also a fascinating diagram (Brett-Jones, 1978: 23) which is entitled, 'What sort of profession do we want?' which indicates that the profession is looking for 'a fair share of the "top" people', which may mean what I call 'the right type' in my conceptual model.

Notes

1 Battersby's Presidential address includes a quote from Macaulay's *Lays of Ancient Rome* which explains beautifully how apparently incompatible men from different subuniverses in the surveying profession can work as a fraternal team,

'Then none was for a party
Then all were for the state
And great men helped the poor
And the poor men loved the great
Then lands were fairly portioned

Then spoils were fairly spoiled
The Romans were like brothers
In the brave days of old'.

2 Interestingly, Orchard-Lisle, (1985: 594) states in his Presidential address that 'perhaps the most telling remark was made to me by a leading socialist: "we are both opposed to free enterprise"'.

Chapter five

Women's presence 1945 onwards

The post-war period

Very little appears to be happening as regards women and surveying in the immediate post-war period. From time to time the journals included references to women secretaries who had retired or who had put in an extra effort, for example, 'Mrs Montgomery has done it again', (*Chartered Surveyor*, January 1962, Vol. 94, No. 7: 384). The only other references to women included the rare event of the announcement of a new female member (*c/s*,[1] March 1949, Vol. XXVIII, Part IX: 507) and the birth of another child to the Queen (*c/s*, April 1960, Vol. 92, No. 10: 518) who, incidentally, must be the most mentioned woman in the journals. About every ten years a woman housing manager wrote an article (Alford, 1954; Ward, 1963; Heath, 1963) reinforcing woman's natural role. However, warning bells were now sounding about articles which did not mention women but whose suggested policies were to have implications for women in the future. An article on shopping (Fenton-Jones, 1962: 79) contains the prophetic quote (innocuous but full of 'power') 'as our car-borne and refrigerator-owning democracy becomes a reality, the chain store and supermarket will increase their percentage of total trade. As a result ... the housewife will shop less frequently'. It describes the expected future growth of what nowadays we would call out-of-town centres and shopping malls. 'Men **are** interested in shopping really', but from a different perspective; indeed, they play a major role in defining the spatial parameters of the social construction of the housewife. Unlike some 'socialist planners', 'commercial surveyors' take shopping, that is, 'consumption', very seriously.

The only other women who appeared in the journals were those from the government who might be seen as having the status of honorary men. Dame Evelyn Sharp was elected an honorary member in 1966 (*c/s*, April 1966, Vol. 98, No. 10: 516) and Barbara Castle, then Minister of Transport (which was amazing in itself), wrote the first non-housing article by a woman in the history of the journal; which was on urban transport

(Castle, 1967). Ordinary women and children are seldom visible within the journals with some exceptions (*c/s*, February 1965, Vol. 97, No. 8: 393; March 1967, Vol. 99, No. 9: 476) in which they are shown as council tenants. Contrast this with the untypical 'human interest' photograph on the cover of the July 1968 issue of two Chelsea pensioners and a 'nicely dressed' boy and girl, standing in front of the RICS headquarters (Vol. 101, No. 1). Another matter of great concern to many women was the proposal for turnstiles in women's lavatories in 1963 (Vallance, 1979: 89). It is commendable that this issue got a mention, albeit non-committed either way (*c/s*, October 1963, Vol. 96, No. 4: 191). The provision of public conveniences has continued to be seen as a serious town planning issue by urban feminists.

The profession continued to grow and prosper in the 1960s in response to the property boom, but this did not yet require the recruitment of women. There were enough grammar school boy entrants. I had intended to talk to women who were representative of those entering surveying within each decade of the present century, but found this presented difficulties. Although I had contacts for the early years, and a progressive increase in contacts from the last ten to fifteen years, I found there was a distinct 'valley' from the late 1920s to the 1960s, which no doubt reflects the wider societal situation for women in those years when they were encouraged to put family commitments before career development. However, I have come across a few women who went into the profession in the 1950s. This was the period of the individual 'exceptional woman' entering surveying. It would seem that completely different rules operated for this minority than for the larger numbers of women entering today. I tried to contact one such woman who is now a very senior surveyor. Her male deputy answered the phone and declared to me (no doubt thinking I was a secretary), 'you don't want to talk to her, she's an old dragon'.

The 1960s and the new era

The mid-sixties were the calm before the storm, and all seemed 'normal'. The journal, reporting on the annual conference, stated that 'the ladies attending the conference had a programme of their own. A demonstration of skin care and make-up was also arranged at a large department store in Nottingham' (*c/s*, September 1966, Vol. 99, No. 3: 159). The journals continued to include advertisements such as one in the October 1967 edition (*c/s*, Vol. 100, No. 4: iii) with a female looking provocatively at a theodolite. Unlike some other construction industry journals which wrapped female forms around every building component and piece of machinery that needed advertising, it is to the surveyors'

The historical perspective

credit that such crude sexism in public, within the journals, is not part of the surveying subculture.

The RICS must have been vaguely aware that the situation might be different regarding women entering the profession in other countries both through FIG (1983) (International Federation of Surveyors) and other 'hints' such as the one from a letter from a man in 1964 stating that the Chinese had put a picture of a woman surveyor (or is she an assistant?) on a stamp (*c/s*, June 1964, Vol. 96, No. 12: 619). This no doubt reinforced in men's minds myths about 'commie women', and did more for stamp-collecting than it did for women surveyors.

In 1967 things hot up, possibly as a result of the beginnings of the second wave of feminism and because women housing managers were having to strike back in view of their losing ground as explained in the previous chapter. A letter entitled 'Even brighter girls' put the case for women surveyors and describes current attitudes, 'surveyors nodding and winking at the mere mention of women, and no doubt falling off ladders at the flick of a mini-skirt' (Smith, 1967). Her letter followed a somewhat patronising, yet vaguely favourable article in an earlier issue (Quoin, 1967: 227). This is one of the first articles by a man on women surveyors, which incidentally embodied the 'tolerant but not enthusiastic' stance to be adopted by surveyors in the following years regarding the entrance of women into the profession: that is until the present *volte face*, brought on by the 'man'power crisis.

Meanwhile, a few women were beginning to enter the new college surveying courses. A letter (Hilland, 1969: 619) expressed concern about 'vocational dead ends being foisted on boys and *girls*' and the need to develop a future officer core within the profession (presumably drawn from both boys and girls). An interesting aside is that with the creation of new town planning courses at the same time, it was noticeable that several surveyors' daughters took this option instead, although it can hardly be seen as more feminine. I have also met several women whose fathers were surveyors and whose brothers became surveyors whilst they went off to do arts degrees, but their daughters are now entering surveying. 'It skipped a generation', as one woman put it, the ultimate in deferred gratification.

At a time when less than 0.5 per cent of surveyors were women, and men appeared to play down the whole issue as being of no importance, the RICS celebrated its centenary, and produced a film of a day in the life of an average surveyor in 1968 (*c/s*, June 1968, Centenary Issue, Vol. 100, No. 12: 633). The film features five partners of a surveying firm including one woman housing manager who apparently was portrayed as an equal partner within the practice, that is, 20 per cent women, as if it were the most normal thing in the world (echoes of Goffman, 1969). Meanwhile in the profession, the emphasis on quantification and

Women's presence 1945 onwards

the growing use of mathematical models reduced the likelihood of women's issues being taken into account in policy-making as they were 'not objective enough'. For example, one article on shopping is based on a discussion of 'user requirements', the users being defined as developers, tenants, retailers, and distributors, but not shoppers (Edgson, 1969: 378). These examples reflect the compartmentalisation of life to which some men are so prone and which continues today. They can be aware of women's rights in one area, such as employment and entrance to the profession, but at the same time never think of women in their policies and professional decisions; this is an important sensitising concept that was to be observed many times.

A more realistic view of women in the RICS was shown in a special edition at the end of the decade, which (unusually) contained pages of photographs of women, who were the office staff who kept the whole edifice running (*c/s*, July 1970, Vol. 103, No. 1).

The 1970s

A letter appeared in the *Chartered Surveyor* in 1973 entitled 'Plumbing the depths' (Ellis, 1973) which describes a historic watershed for women. Separate lavatories for female members were introduced at the RICS headquarters. Space matters. Many women impressed on me the importance of this event. Then in 1976, for the first time, the new President of the RICS referred to women in his Presidential address, asking why there were not more women in the profession 'in these days of equality' (Franklin, 1976), apparently almost giving the impression it was the women's fault. This led to a series of lively letters from women in subsequent issues. For example, one wrote a letter on women in the profession, (Snow, 1977), and the JO (Junior Organisation) news (Allen, 1977) featured a sympathetic item from a sensible man. Some of the letters were even quite feminist but not radically so (Sousby, 1977). I do not consider that the age of the bourgeois feminist had quite crystallised, as it was to ten years later within the climate of the enterprise culture. Women surveyors, then, were more likely to espouse a vague liberal feminism, or 'we simply didn't think about that sort of thing'. After years of silence on the matter it was commendable that the journal was willing to publish such letters, something that neither the planners nor the architects were to do for many years. However, judging by the titles given to some of the women's articles and letters (as will be illustrated below) some no doubt saw such items as the source of titillation, rather than as serious matters for the consideration of the profession.

There was by now a very small number of women entering the colleges and practice. Such women were encouraged to believe that they weren't any different, and that it was the same and fair for all. Women were likely to be told that they were very lucky to be there, and that everything was wonderful. I well remember being told this as a town planning student and was made to feel really inadequate and 'immature' for expressing dissatisfaction. Some women dropped out. Nowadays, young women look back on the sixties as the Dark Ages and one is encouraged to say how bad it was. In the seventies, many men surveyors saw individual women surveyors and planners as a great 'oddity' or 'novelty' and belied their feelings by putting into action 'the token woman approach'. One young woman surveyor, who confessed that she was not 'women's lib' but that her surveyor father was (Vol. 111, No. 11: 78, October 1978), was elected as the first woman chairman of the JO. She then produced a chatty column in diary form in the journal throughout the year 1978–9. She was sent on a grand public relations tour of all the branches. Her diary included descriptions of what she did, who she met, and how she was wined and dined. Many surveyors have commented that no other JO chair before or since did anything quite like this, although many admired her for being a good sport. The end of her term of office was marked with an article in July 1979 (c/s, Vol. 111, No. 12: 24 of Supplement) in which she confessed that now she would have more time to concentrate on one of her favourite hobbies, 'watching cricket'.

Years later, the diary was seen as just 'a bit of froth', by those involved and 'frankly best forgotten'. It often takes several years to see the real significance of such events and 'opportunities' from a feminist perspective. Meanwhile, outside the profession, the women's movement was growing, but it was only gradually penetrating the landed professions; indeed, there was scarcely a ripple in the private sector. There was also very little British urban feminist literature until the 1980s, as indicated in Chapter 2, although various women, including myself, were 'thinking and worrying' about it all on the quiet, ready to start writing about it (*Built Environment*, 1984; Greed, 1984a and b). However, those women working in local authorities, particularly in the Greater London Council and other more 'progressive' authorities, were inevitably aware of the beginnings of the establishment of women's committees which made their views heard on urban issues such as housing, public transport, and safety.

The 1980s

As the profession expanded in response to the enterprise culture of the eighties, it became necessary to bring in more women rather than recruit men of another social class. The journal actually states that in 1980 the percentage of female membership had reached 1 per cent, representing a 100 per cent increase over twenty years (c/s, April 1980, Vol. 112, No.

9: 8 of Supplement) whereas in 1989 it was over 3 per cent (excluding student members), representing a considerable acceleration in the rate of growth over the last nine to ten years. In comparison, there were 12,690 fully qualified surveyors in 1955 (Thompson, 1968: 349), 18,290 in 1966 (Thompson, 1968: 341), 40,818 in 1979 (SITE Report, 1980: 34), approximately 56,910 on the last day of 1986 (RICS, Records Office), and nearly 60,000 in 1989 (Appendix 1). This also represents considerable growth, but not at the same rate as the growth of women within the profession.

By 1986 there were 20,666 students and probationers as against the 56,910 full members, of whom 13,644 were under the age of 33 (RICS, Records Office, 23.12.86). There were 2,745 females of all categories (students and full members) at the end of 1985 which rose to 3,581 by the end of 1986 and 4,703 (out of a total membership of 81,602) in 1989, illustrating the rate of growth. Of these 4,703, 1,475 were students and 1,370 probationers, leaving 1,847 who were full members (plus eleven 'others') of whom 1,250 were under 33 years of age, that is, only 606 were older women surveyors (Appendix 1). In the past, many women who entered surveying subsequently left owing to problems of child care and lack of career satisfaction, but this trend is now slowing down and more women are determined to work all their lives. There does not seem to be a clear 're-entry' age as such; rather, a gradual build-up of more women working all the way through but some seeking to return (with mixed results) in their late thirties (compare JO, 1986: 13 and Elston, 1980).

More women were going into the more technological areas of surveying. For example, there is an announcement of the existence of the first woman hydrographic surveyor headed 'First sea lady' (as opposed to sea lord) (c/s, June 1980, Vol. 112, No. 11: 9 of Supplement). I was able to interview her and found that this small announcement belied a lifetime of personal determination and a most interesting and unusual series of fortunate 'breaks' which enabled her to get into what is an almost womanless area. (Many women surveyors believe strongly in the serendipity factor in their lives.)

Other women 'appeared to appear' in photographs in quite macho, 'hard-hat' roles, but such apparitions are not always trustworthy images of the true situation (c/s, 21.6.84, Vol. 7, No. 12: 822). I suspect that some men find the macho woman a 'turn on', although many feminists for quite different reasons also believe in the importance of women in technology. About this time, the 'woman engineer syndrome' became almost a cult in its popularity, but whether it really achieved anything for women in society is another question. Also, the first woman chairman of a regional branch of the Rural Land Agency Division is announced (c/s, 7.6.84, Vol. 7, No. 10: 676) one of several 'first women'. As one such woman told me, 'they only put the first one in, they don't bother after that, especially if the women continue to outdo the men'. However, these appointments at least established to other women alternative role models within the profession in areas and at levels where women had not previously entered. Such women often had great difficulties and obstacles put in their way in attaining these achievements, but once they succeed, it is often presented as if the men were on their side all the time (but as will be seen, in fairness some men do encourage women in the more technological areas).

The Lionesses (an unfortunate name which led to endless comments about 'red meat' etc.), which later became known simply as the women surveyors' association, was established at the start of the 1980s. Items appeared concerning the creation of this group and debating which ladies are entitled to wear the badge introduced by the Lionesses for women surveyors. It appears some surveyors' wives and girl-friends were wearing it, having been given it by men surveyors. This was seen as an affront to the hard-earned status of women surveyors (c/s, August 1980, Vol. 113, No. 1: 3; September 1980, Vol. 113, No. 2: 130). Following this, quite a strong article appeared on 'what it's really like to be a woman surveyor' (Smyth, 1980) which was very different from the traditional 'how lovely it is to be a woman housing manager' article of the past. This was followed by a letter from a woman district valuer concerning men surveyors' attitudes to women, which actually stated that the writer had found one of the past presidents 'patronising' towards women (c/s, February 1981, Vol. 113: 505).

There was considerable discontent among some of the women who were 'getting older' mainly because the profession had 'allowed women to enter' without, for its part, changing anything to accommodate women who wished to be both women and surveyors at the same time and have children. However, women 'put a brave face on it', and actually appeared to be rising within the profession and entering previously male-only preserves. Whether this is a tribute to the profession or the result of women working twice as hard is another matter which will be discussed in later chapters. Women began to get more official recognition and posts both at Junior Organisation level (c/s, March 1982, Vol. 114: 478), and in a few of the erstwhile exclusively male committees within the RICS itself. Today there are three women committee members out of a total council membership of 108, which is almost the same proportion (2.7 per cent) as women's representation in the profession as a whole (3.1 per cent) (e/t, 14.7.89, No. 1003: 12). It is fascinating that more women were beginning to appear as writers in the journals. This began in the late seventies with women surveyors writing articles on professional matters (e.g. c/s, March 1979, Vol. 111, No. 8: 302). More surprising, women journalists who were not surveyors began to take over reporting highly technical issues. A glance at any of the issues of the journal today will show the frequency of the work of one such woman who became Building Correspondent, and another who frequently produced articles on legal matters both in the *Chartered Surveyor* and the *Estates Times*. In 1983 a woman surveyor became Assistant Editor and others are following in her footsteps (*c/s*, 27.1.83, Vol. 2, No. 4: 211).

These examples illustrate that, for a woman, professional knowledge and expertise does not ensure power and partnership; rather, they may find that it enforces their feminine role as researcher or writer within the profession, as 'helpmeet' (Heine, 1987), although some are quite influential in this role. Indeed, there are examples of women in the new role as writer, writing about other women in the traditional role as housing expert (Hillel, 1986; Coleman, 1985). Even women who have entered the more male technological areas may find that they are still expected to perform female roles. A woman building surveyor is commissioned to study buildings for the elderly. These are all examples of the channelling of women into specific roles within the world of surveying.

Meanwhile, articles written by men continue the long standing tradition of adding an element of 'harmless innuendo' to otherwise non-sexed matters such as planning law (*c/s*, 9.6.83, Vol. 3, No. 9: 559): 'it attracts the men's attention as they wouldn't read it otherwise'. Sometimes it would seem that the importance of an issue can be judged by the level of innuendo used in presenting it. Other items in the journals at this time give one the feeling that the profession is still 'truly a league of gentlemen' (*c/s*, 17.3.83, Vol. 2, No. 11: 585). It is also noticeable that as women seek to raise serious issues themselves, or announcements are made about them, they are often titled in a way that trivialises the issue. Matters of sex discrimination are reported under headings such as, 'estate agents who preferred men' (*c/s*, 22.9.83, Vol. 4, No. 12: 591). However, they do at least give space to these issues. The journal even admits the predominantly masculine nature of surveying in reviewing a new book on careers in quantity surveying (*c/s*, 1.9.83, Vol. 4, No. 9: 447) although the male author is to be commended because he purposely used women surveyors for a third of his examples of people with successful careers in surveying. Or is he? With the minute number of women in quantity surveying, it might give women a misleading impression.

Whilst some men are aware of the issues, others continue as 'normal'. For example, Miss World (*c/s*, 19.1.84, Vol. 6, No. 3: 168 and 113) was invited to the opening of a new office development in Bedford (this being only one example among many of a continuing obsession by some men in the property world for such women). In fact, some surveyors may not see women surveyors as being of the same species.

The historical perspective

Comments such as, 'let's go and get some real women' can still be heard when men want to neuter women colleagues publicly. However, there are more positive and realistic role models which appear from time to time in the journal such as a magnificent feature on a husband and wife team of surveyors in a high prestige practice (Whelan, 1984).

It was becoming fashionable to be a woman surveyor. *The Archers* radio series made one of its young women characters, Shula, a surveying student and, as a gesture of good will, the actress was invited to the RICS stand at a major building exhibition (*c/s*, 12.4.85, Vol. 11, No. 2: 124). However, at the same time, a story about an all-women practice is covered in innuendo, 'ladies take off' (7.3.85, Vol. 10, No. 9: 633) and 'ladies uncovered' (4.4.85, .Vol. 11, No. 11: 4). Complaints are even received that Landseer, a weekly feature, is too sexist (16.5.85, Vol. 11, No. 7: 456), but one gets the impression that such accusations are seen as a joke. Everything seems to be written for an assumed male audience, and women carry on a serious discussion with each other through the vehicle of the journals in spite of the ongoing trivialisation of the issues they seek to discuss. Nevertheless, particular problems for women such as doing CPD (continuing professional development) whilst bringing up a family or moving around with a husband are discussed, albeit under headings such as 'sex and the single surveyor' (20.6.85, Vol. 11, No. 12: 848), this being presented alongside a letter from a sympathetic man, the word 'feminism' being used for the first time in the journal.

'Candidate urges Lionesses to bite back' (*c/s*, 8.5.86, Vol. 15, No. 6: 508) is another serious item which is trivialised by its title. Nevertheless, some good articles and letters find their way into the journal. In 1986 the problems of combining child-rearing and surveying are discussed for the first time (Turner, 1986), and the problems of women working (Marwick, 1986) are aired further. Coming up to the present and following the case of Suzy Lamplugh, a woman estate agent who went 'missing', it would appear that the position of women in surveying is taken far more seriously. A whole article is devoted to the safety of women in the profession (Cornes and Lamplugh, 1987). In particular, Suzy's mother, Diana Lamplugh, has done much to raise the profession's consciousness of such issues, and to help women deal with the problems they are likely to experience by giving an extensive programme of talks and courses throughout the country.

However, some women feel that 'concern' by men (whilst apparently genuine and serious) might be a mixed blessing as it reinforces the idea that women need special attention and cannot be sent to all the places that men can go, that is, women, not their assailants, are the problem. In fact it might be history repeating itself, as women have been kept out of the urban arena in the past by being told it's too dangerous (i.e. male) for them. This might have the wider effect of excluding them from certain policy-making areas too.

Challenges to success

The profession has continued to grow in status and numbers in response to the buoyant property market, in which landed capital had almost replaced the importance of industrial capital as a primary component of the economy. This dynamic phase of professional development is prefigured in the tone of the SITE Report (1980). Although many surveyors imagined that surveying was now truly a higher profession, even drawing frequent analogies between the medical profession (Hanson, 1983) and surveyors as 'development doctors', others continued to express concern about the standards of surveying education ('RICS comes bottom of the class', *e/t*, 13.6.86, No. 849: 9). Whilst the enterprise culture of the 1980s has undoubtedly placed young chartered surveyors in among the membership of the affluent yuppie culture, at the same time the deregulation of financial services and the threat of competition from other professionals has caused older and wiser surveyors to regroup and defend themselves (RICS, 1987b). It is significant that the annual conference in 1986 was entitled, 'Professions in Crisis: New Opportunities for Chartered Surveyors' (RICS, 1986b). Also, whilst surveying education has increased tremendously, government cutbacks in education and falling birth rates have given surveyors much cause to worry (Cox, 1985).

If numbers drop, what happens to all those educational and professional empires that are built on these figures? 'Any significant increase in participation rates must mean a disproportionate increase in participation by one or more of the groups in society identified in the National Advisory Body Strategy Advice of 1984 – *women*, mature students and those without standard entry requirements' (*NAB Bulletin*, Autumn, 1987: 23, para. 9). There are other changes afoot in surveying education which at face value might appear to make it 'easier' and more accessible for women, such as proposals for the encouragement of greater 'access' at different levels (NCVQ, 1987) and ideas for restructuring courses on a modular basis. Also, the old RICS Parts I, II, and III external examinations are being phased out in conjunction with the above changes and replaced by a range of in-college part-time routes and distance learning packages. Time will tell as to whether women win or lose as a result of these various current proposals. Some are worried whether modularisation, and also distance learning (in which courses would be divided into subject units with more emphasis on self-teaching materials), will make courses more impersonal. This may make it more difficult for women without any technological background to understand the nature of subjects of which they have no prior knowledge and which they need 'explaining'. Many are concerned that the 'glue' will be missing, or at least less in evidence, that is, that in the drive for efficiency there will be less space for communicating and absorbing the essential subcultural characteristics of surveying. This may not matter if what is required is 'obvious' to students because they have been 'precisely' pre-socialised as to what surveying is all about, before they come on the course, but many women, in particular, have not.

This chapter is relatively short but it is hoped that the reader herself, especially if she is a surveyor, will add details of future developments as the numbers and influence of women within surveying increase. There is now a new generation of women entering surveying who have grown up under the influence of feminist ideas and reforms within the general milieu of society, who cannot see 'what all the fuss is about'. Some younger women appear very confident, and seem to take the apparent 'equality' for granted, perhaps not realising what has gone before. Older women are more cautious. It is to be hoped that this time round we are experiencing real change and not another 'false dawn' to be followed by decline (Greed, 1988: 196). Indeed the RICS does seem to be taking the issue more seriously now, as manifested by a recent series of articles in the *Chartered Surveyor Weekly* (*c/s* 22.2.90, Vol. 30, No. 8: 108; 1.3.90, Vol. 30, No. 9: 140; 8.3.90, Vol. 30, No. 10: 142; 15.3.90, Vol. 30, No. 11: 131; 22.3.90, Vol. 30, No. 12: 5 (Editorial); and ensuing correspondence on letters pages.

Note

1   In this chapter the *Chartered Surveyor* has been abbreviated to *c/s* and the *Estates Times* has been abbreviated to *e/t*.

Part three

Education and practice today

Chapter six
The educational context

Apologia

Chapters 6–9 comprise the most 'sensitive' part of the book, as they record what women and men surveyors have felt, and said, about the position of women in surveying education and practice. In reading these chapters, it is suggested that the reader refers to the model (Figure 1, page 21) as she goes along and makes her own connections between the levels and columns from the illustrations given. This will help her understand 'how' the values of the surveying subculture affect the position of women within the profession and the lives of women in society as consumers of the built environment. The purpose of this present chapter is to discuss the nature of present-day surveying education *vis-à-vis* women, incorporating ethnographic insights where appropriate.

Surveying is not especially 'bad', as a similar situation pertains in education and practice in the fields of architecture and town planning; indeed, many women believe that surveying is actually 'better' than these areas. However, the findings might surprise some surveyors, although I believe the profession is resilient enough to take it and is seeking to improve the situation. At a time of 'man'power shortages, we must do all we can to enable, and not deter, women to participate fully in the profession; indeed, the economic and efficient use of 'human' resources through enlightened and sensitive management actually saves money, that is, it reduces the cost factor at all stages of the development process.

Methods and reactions

First, I wrote to the majority of colleges that teach surveying degree courses which, as stated, normally are of three years duration with some courses having, in addition, a 'sandwich' component which may range from a few months to a year depending on the specialism (Whittington,

1987). Following this, graduates must complete the TPC (test of professional competence) by two years of supervised training and experience followed by a practical test. The total numbers on the mainstream full-time courses average about fifty students per specialist year group, with much smaller numbers of around ten to fifteen in some of the more esoteric areas (and some part-time routes), and a few instances of 150–200 students in combined first-year groups in certain colleges. I asked for information on the numbers of women students past and present, their levels of examination achievement, and (where appropriate) their subsequent career progress, and also invited general comments on the topic. I received twenty-two useful replies in response to approximately thirty-five letters. Most of these contained specific data ranging from minimal five-line tables to sheaves of information. A few colleges requested that the source of material should not be identified. As stated, I decided not to include any detailed figures, names, or sources because of the potential sensitivity of the situation; however, I will describe the overall national situation as revealed by this data.

Some replies were extremely enlightening (not always in the way intended) and reflected the values of certain aspects of the surveying subculture. For example, some course leaders did not send me any material on courses on which there were very few women, having decided for me that 'there were so few as to be insignificant' (echoes of Becker *et al.*, 1961: 60). Other departments seemed embarrassed about their low female numbers and wanted to know what the going rate was before divulging their figures. Once they were reassured that they were 'all right', they might even boast that their numbers were slightly higher than the average, almost as if it were a competition. Some course administrators had a misdirected equal opportunities policy of not recording the gender or the first name of students, and the infamous preface 'Miss' was no longer used, thus rendering the women invisible and therefore difficult to count. The emphasis on treating everyone the 'same' somewhat undermined my efforts. It created an atmosphere in which one felt awkward, in drawing attention to what was portrayed as an unimportant issue. Even the Council for National Academic Awards had admitted that 'the question of gender has not been a matter of comment for the [Surveying] Board' (letter 20.11.85), although it produced a range of other fascinating figures.

One male course leader told me scathingly, 'I've really no idea off the top of my head how many women we have in the department'. Other men were extremely helpful and well-intentioned, especially, significantly, those in the technological areas of surveying. It is fascinating that some men react one way and others another, paralleling the fact that some women become feminist, others do not, and some, sadly, can best be described as 'patriarchal women', showing again that there are considerable differences in attitudes within the gender groups. However, I could not find a consistent 'badness' to 'prove' patriarchy existed, but I could validate a 'feeling' of it, based on the very fact that the situation proved so patchy and unpredictable. The **capricious** way in which women were treated, sometimes well, sometimes less well, with no apparent pattern to it, was unsettling in itself. As one woman put it, 'you never know what to expect, you can never relax'.

Second, I visited a representative range of colleges, including a polytechnic in the Midlands, another further north, a major southern university, a London polytechnic and a provincial technical college in the south-east, meeting groups of women students. I also talked to a range of course leaders, lecturers, and ex-students, both male and female, in groups and individually, from a variety of other colleges, in some cases on a face-to-face basis, and also through the extensive use of telephone conversations as part of my dispersed ethnography (as discussed in Greed, 1990b). Whilst conversing with people, I sought to ascertain their views on the four substantive issues identified in Chapter 1 (briefly: surveyors' world view, and their assumptions about land use, other people and women, and who is the 'right type') as well as discussing with them their experience of surveying education.

Growth of courses

As will be seen, the increase in the numbers of women on courses went hand in hand with the expansion of surveying education itself. Over the last twenty years, diplomas were gradually changed into degrees, and ordinary degrees were turned into honours, and overall the emphasis has moved towards more surveyors qualifying by internal courses, surveying being smitten in moderation by the diploma disease (Dore, 1976). This increased credentialisation (Collins, 1979), far from excluding women, actually worked in their favour; indeed, the attainment of competitive academic entry requirements gave women more 'right' to enter. Colleges welcomed 'brighter women' who, I was frequently told, were 'good for the chaps; they make them work harder and behave better' (parallels with 'mixing' schools). Many women ex-students were of the opinion that they had been expected to bear the burden of increasing the standard on courses. Interestingly, they believed, 'thicker men' were accepted with lower grades, 'the true mark of equality will be when we get the thicko female being accepted on surveying courses'.

The numbers of women on courses grew very gradually. Taking a typical general surveying course based on a cohort of, say, fifty students, I observed what I call the 'splutter effect'. A few sparks start the fire. It gradually glows then dies down and suddenly comes to life again, and then consistently burns away more strongly. Back in the late 1960s, there were seldom more than one or two women on a course, and in some years none at all. Then the numbers went up to say three, of which perhaps one would leave the course after the first year. Old-timers in departments would explain, 'one was too attractive, and the other was too bright' or 'she left to get married; she lost interest'. By the mid-1970s there might be, say, three or four women, one of whom would be likely to be an overseas student. The lack of women at the beginning led to unusual alliances being formed. One woman surveyor, who had been the only woman among 100 male students in the late '60s, said she used to go around with the two male black overseas students, 'we stuck together because we all stood out and were different'.

Then the figures might fall again, and then a few years later four or five women might appear, then none again. The figures would hover around the four mark, until the early 1980s as the fire took a long time to draw. By the mid-'80s there was an observable increase, a catching fire, with perhaps six or seven women per cohort. At the same time, the size of the classes as a whole had increased, the unit of fifty having risen to nearer sixty on the most popular courses. Notwithstanding women's part in this and some men course leaders' attempts to recruit fairly, other factors were mentioned in some colleges such as the need 'to recruit more women to allow for future student cutbacks' and 'to protect the course'. Also, in the less popular specialisms of surveying, there is some competition between courses to attract more students, and so they are interested in recruiting women to 'keep afloat', especially overseas women who might in some cases be seen as being 'worth more' or having a greater inclination towards the technological areas than indigenous women. Paradoxically, it would also seem that the greater the popularity of a particular course, the more women there will be (who are generally seen as being 'brighter' and 'classier' than the men).

Distribution of women

The representation of women in the different areas of surveying education will now be discussed in a similar sequence to the summary of the range of surveying courses (RICS, 1987a), given in Appendix 3. As a general principle, the most technological courses have the lowest numbers of women, and the most socially-oriented courses the highest numbers, with the more commercially-oriented courses being about in the middle. Starting with the most technological, courses such as surveying science, geodesy, photogrammetry, land surveying, etc. are concerned with 'real surveying' of land and sea (and nowadays space). These areas make up the smallest divisions of the RICS, with the smallest numbers of courses. However, because of recruitment policies in the few colleges offering these specialisms, such as a leading one in a London polytechnic, percentages of women students on these courses are now much higher (10 per cent in some instances) than they were (around 2 per cent), but actual numbers are very low relative to other specialisms (Appendix 3).

As indicated in the historical chapters the status of these specialisms within the surveying subculture has changed over the years. Some land surveyors do not even belong to the RICS, aligning themselves either with land surveying bodies, geologists, or physical geographers. Indeed, one male land surveyor referred to chartered surveyors as 'just a load of estate agents'. However, the whole profession takes its ethos from 'real surveying' which sometimes 'puts women off'.

Whilst some see it as a sign of equality and progress for a woman to go into a technological professional area, it may be a dead end. As one land surveyor explained to me, 'the countryside has all been mapped, nowadays they are more likely to be asked to survey changes to street layouts in urban areas'. Land survey is arguably a declining area (except for the new high technology areas), and it has no social policy content for those who seek to change the environment, as against recording it. Technological areas appear powerful because they are so male-dominated, but the greater power may lie in the office-based areas of general practice surveying. Interestingly, several women in these areas have given me glowing accounts of the help that men lecturers gave them. Different mechanisms operate in different specialisms. One woman on her own in a technological area may paradoxically receive better treatment (in an otherwise highly chauvinistic subuniverse) than say ten women (i.e. a threat) on a mainstream course.

Minerals surveying scores the lowest of all with around twelve women, of whom six are students, five probationers, and at least one a fully qualified surveyor (Appendix 1). Such women are 'different' from the estate management women, and this again illustrates how diverse women surveyors are. Further along the course spectrum is building surveying with women making up 2–7 per cent of students in colleges with slightly higher numbers being found in first-year groups. This area is often written off by women as it is perceived by them as somewhat down-market, lower class, and full of 'rough beer swilling yobos'. In reality, 'building surveying' can cover a multitude of professional activities of different academic levels, some of which are high status, and some admittedly lower. Quantity surveying courses nowadays attract 3–10 per cent of women students after years of very low numbers, but there may be more overseas women than indigenous women, 'every course seems to get its four Far Eastern or Arab women'. This is not necessarily a sign of equality but often the result of a non-gendered quota system on the part of the sending country. 'Class' (and 'caste') can cancel out both gender and ethnic origin as high caste overseas women (even Middle Eastern women wearing traditional Muslim clothing) have been known to take over as dominant leaders in predominantly white macho-male groups.

In the middle of the spectrum, around 20–25 per cent of students on general surveying and estate management courses are women, even more in some London polytechnics (Whittington, 1987). Taking all specialisms together, there is approximately a 15 per cent average of women students on surveying courses (RICS, 1989), in comparison with 9 per cent for engineering courses, around 20 per cent for architecture, and up to 30 per cent for planning in the current intakes (Healey and Hillier, 1988).

At the other end of the course spectrum are the areas with the highest percentage of women, namely housing undergraduate courses where up to 50 per cent are women. These are courses where one is also likely to find more working-class and ethnic students (perhaps 5 per cent), leading a male estate management student in one college to make the memorable comment, 'they're all wogs and women on the housing course'. This is by no means an isolated incident, and reflects a certain 'tendency' within the subculture. In fairness, other male students would not dream of making such comments and are more responsible. Men surveyors have told me, apologetically, that such examples are sometimes caused by embarrassment rather than actual antagonism, reflecting the views of 'callow youth' (not that that means it hurts any less nor is it any excuse). However, one can also find traces of such attitudes within the profession itself in the past concerning 'the coloured man' (*sic*) (Langdon, 1949: 203) and in the present, as will be illustrated in later chapters. Nevertheless, students from several colleges have observed that, 'if you're not good enough to do surveying, you do housing' and even, 'I didn't realise it was part of surveying'. (In fact 'housing' is rapidly becoming a university postgraduate growth area.) A course's standard may be assessed by its 'class' rather than by its academic quality.

There are also other 'levels' to surveying education. Joseph commented that surveying may be seen as a safety net for the professional classes (Joseph, 1980). Nowadays, this is not particularly true of degree courses where a commendably high standard is reached, but there is a range of diploma and 'less academic' courses where it might apply, where one 'A' level entry is more common, as against the three 'A' levels required for many of the degree courses. Such courses either give exemption to non-RICS qualifying bodies, or higher technician grades, or in some cases are a bridge to other qualifying courses. They may contain very mixed ability students, ranging from those that are seen as 'simply thick' (by degree students and some lecturers) and have got bad grades in spite of attending 'good' schools, through to those who are bright and have been to 'bad' state schools, and those who are intelligent but are 'practical rather than academic' and 'know exactly what they want to do'. In particular, one finds young men who have already been out in practice for a couple of years in estate agency. Overall the girls might be thought to be 'a little brighter' than the average boys on such courses (and are probably there for somewhat different reasons) although it has been commented by various people that 'one gets some really dim females too' (the true mark of equality?).

There are also certain northern technical colleges where it is possible to do courses that give RICS exemption. Such courses may contain few northerners or women, but a surprising number of southern ex-public school boys. Many bright northerners (mistakenly?) head south, as they are convinced that attendance at a southern college is a stepping-stone to a job in London.

There is another level still, that of surveying technician courses, which are both more 'male' and more 'working-class' than any of the above courses, but I would argue that many of the students are quite as bright but magically, 'they know their place'. Also on the edge of surveying education, there are occasional evening courses for 'negotiators', that is, people who sell houses in estate agency (realtors). These are often held in conjunction with commercial companies, attendance being rewarded with a certificate or even a modest promotion. Such classes usually consist of 90 per cent women who come from diverse backgrounds, but they are often assumed to be of low academic ability. However, lecturers may find that such women are, in fact, exceptionally keen and 'quick' compared with the more ponderous pace of some full-time male students.

Full-time courses are only part of surveying education. There is still a sizeable number of people qualifying by external, part-time, and correspondence routes. The profession has not only expanded but also created distinct levels and status groups within itself, with people arriving at their different destinations within the profession via different college routes. There is a great difference between a part-time building surveying student employed by a local authority and doing a day release course at a provincial technical college and a young chap going to a prestigious full-time university course; but they are both going to be surveyors.

Part-time and correspondence course students might include the sons of high status surveyors who 'can see through education' and have no desire to indulge in anything as down-market as being a student (compare *Chartered Surveyor*, 23.2.84, Vol. 6, No. 8: 496). There is also a substantial number of non-cognate (non-surveying) graduates on part-time courses, creating a very mixed-ability class altogether.

Surprisingly, there are very few postgraduate surveying degrees which again reflects the anti-academic ethos (as described by Joseph, 1980); but some see this as the next growth area for surveying education. In contrast there are around twenty-five colleges offering town planning degrees, half of which are universities, offering between them twenty-seven postgraduate and sixteen undergraduate courses. Therefore town planning is of higher academic status, but is generally seen as lower status by surveyors. Again 'class' and academic status are apparently not related.

The women on part-time and correspondence courses in surveying are usually in a minority, as it is less common for women to study and work at the same time (although they may comprise nearly half of students on housing part-time courses). In the past they seemed to come in two different 'types', both of which were fairly 'unusual'. First, there were those with a degree in a non-cognate subject, sometimes from Oxbridge, who then qualified as surveyors at the local technical college, 'it was quite a come-down but one had to be very philosophical about it. Some of the lecturers were very polite almost as if they were afraid of me'...'I had an arts background in English, and had never had to memorise so many facts before, all these boys straight from school were so much better at it'. Second, there used to be a very small number of women who, having gone straight from school to work in a local authority, were sent to technical college on a day release basis, 'they put me on the surveying course because there was no other course available, I don't think they, or I, had any idea that I would end up as a senior chartered surveyor'. Delving into the past and talking to some now elderly male lecturers, it would seem that such 'girls' were not taken very seriously, 'I can't imagine what they came on the course for'. However, I found that several such women had risen way beyond some of the men in the same classes. After all, a woman had to be fairly exceptional to attend a run of the mill course.

Several men and some women surveyors have commented that, until the recent influx of greater numbers, women surveyors have fallen into a range of two or more extremes and types, somehow missing the range of 'ordinary women' in the middle, whilst the men seem to be drawn from a much narrower range of types. But one of the paradoxes of my research findings was that some quite exceptional women really believed they were ordinary and could not see what all the fuss was about.

Women lecturers

The percentage of women lecturers varies considerably from college to college, from about 1 per cent up to 10 per cent, with a few exceptional colleges with around 20 per cent. It is impossible to generalise, as I could find at least one woman lecturing in just about every subject, although often she might be a part-timer or servicing from another department; and of course the situation changed from term to term. It is not like in geography where there are clearly identifiable departments with no women lecturers (McDowell and Peake, 1989). However, there do seem to be more full-time women teaching valuations and town planning than other subjects, the former having a zero social element, and the latter not necessarily being taught from a socially aware perspective but as 'developers' planning'. One also has to be careful to ascertain what exactly women in the more technological areas are actually teaching. For example, women in quantity surveying may be teaching building contract law rather than construction. The few women I came across in the technological areas seemed to be over-qualified, such as women architects teaching basic construction courses, or even draughtsmanship. Likewise, there is quite a cottage industry of women lawyers and other highly qualified women professionals taking on marking for surveying correspondence courses whilst 'indisposed'.

Several younger women were of the opinion that older women in surveying, who had entered the profession when there were fewer women in it, were likely to be of a higher calibre (and some said 'totally different class') from equivalent men in surveying education. It is therefore not surprising that at least three women in surveying education are at head of department level, or is it? Indeed, proportionate to their minute numbers, women are at more senior positions in surveying departments than other departments within at least two polytechnics. Women either fit 'precisely' (Joseph, 1978) and, for their numbers, achieve higher levels of responsibility than equivalent men, or they are seen to hold a more radical perspective than the others and are perceived as 'different' but are still accepted as their presence shows that surveying education is progressive (in the same way that bright hard-working marginal men were welcomed in the past). This is a classic example of women who may be operating from a feminist or 'different' perspective, unwittingly fulfilling the requirements of the patriarchal subculture. Again, it seems that there are two extremes without many in the middle (but some men surveyors believe there are now 'more ordinary women around too'). It is a telling question, which one 'feminist' woman surveying lecturer in a university department told me I must put in, that one of her students had asked her, 'will I have to cut my hair to succeed?' as she had noticed (mistakenly?) that most successful women in surveying education and practice had short hair!

Many surveyors were of the opinion that nowadays, because of the demand for surveyors out in practice, 'only women will be left in surveying education'. Indeed, full-time vacancies are increasingly being filled by younger women in some departments, but not necessarily for that reason (this is the exact opposite of many other areas of higher education where to be an academic is 'the main career' and there is no professional practice career alternative). Also, there is, in just about every department, say, three part-time or associate women lecturers. In fact, women in some colleges believe there is a definite policy of shifting the work onto such women whilst the full-time men retain control and occupy themselves with 'administration' and private consultancy. However, it is quite a trick to retain a male ethos to a department and a profession whilst using women as the mouthpieces for the transmission of the subculture. This is a million miles away from women being in control and teaching from a feminist perspective. Others would argue that many dedicated men, willingly, more than pull their weight (especially since cutbacks have come in) whereas some women with young children were seen as 'highly unreliable'.

Within the anti-academic atmosphere of surveying, the position of the 'teacher' is an enigma. Many students see lecturers as failures because they are not in practice (but women are seen as having an 'excuse'). Students may not necessarily respect lecturers, but they will still strive to reproduce their views in order to pass exams. As Joseph pointed out (1980), there is a dilemma in surveying education. To get degree status, there must be some tolerance of 'academics'. However those that are 'best' are those that are in practice, who are naturally more likely to reflect an entrepreneurial outlook and to lack concern for social issues or education. Alarmingly, some women lecturers were seen as negative role models by younger women, 'I never wanted to be like her, she terrified me', but some were seen as very supportive. The legendary male 'part-timer' (everybody 'knows' one) who spent most of his time in consultancy (and came late, or not at all, to lectures), may have had a very much higher reputation with some men than a conscientious woman part-timer, or full-time male lecturer dedicated to education. In comparison, this is even more so in architecture, where all the best male lecturers are 'part-time', and the women (few though they be) are more likely to be full-time.

Mention must also be made of the 'other' women in surveying, the office staff, catering staff, cleaners, technicians, and librarians. For many years these were the only women in many departments, and students got their image of the 'place' of women from such women's role in the department. Such women seem to take on the ethos of the department. A new typist, who typed out a conference programme for me, couldn't read my writing and typed in 'land' instead of 'lunch' in the midday slot each day, showing that subconsciously she had learnt the most sacred word in the subculture! Before I had my word-processor, the typists had typed up an earlier paper for my research and shared their own valuable views of the situation. If they disagreed with what I wrote, or had some additional insight from their perspective, they would tell me so, which I found very helpful.

School background

Surveying students (presumably male ones) are frequently referred to by town planning students as 'just a load of rugby-playing, public school thickos' but this is not so. I went through several sets of student records of both males and females, checking their schools in the public schools' yearbooks, but still ended up confused by the large number of quasi-state, grammar, and 'not quite what they appeared to be' type schools they had attended. In the post-war period it is estimated that the percentage from state grammar schools rose to 50 per cent reflecting the expansion of surveying by recruitment from more meritocratic sources (Davies, 1972). Many would say that the public school contingent is now rising again in the present generation. A small but significant number of surveyors came from comprehensives, but it is noticeable that the range of sporting and academic opportunities offered in 'their' comprehensives is not the same as that in the average one. Surveyors are very good at knowing what is 'available', and using whatever the current educational structure offers to their best advantage.

I did not make a special investigation of students' social class origins (as had Joseph, 1980) but had a reasonable impression through my ethnography and past experience as a lecturer. As with 'gender', it is simply 'not done' to raise such matters, and it may have been counter-productive. However, it would appear that a considerable number come from professional, entrepreneurial, or farming backgrounds. There were a fair number with mothers (as well as fathers) who were in the professions, including accountants, lawyers, and some town planners (that is, entrepreneurial families with possible bourgeois feminist mothers). Others had mothers who were teachers, and some such women made it very clear that they had come into surveying because they did not want to teach. Others had quite 'ordinary' parents where the mother 'stays at home'. However, exasperated women students may see this 'normality' as a problem, and have been known to exclaim in response to perceived sexist attitudes, 'what can you expect with the sort of family backgrounds they come from?'

Admissions

One of the biggest problems is knowing of the existence of surveying, and knowing what it is. Interestingly, those who do know are more likely to want to go into it because of the high professional status it gives, rather than purely because of an interest in the built environment. Very occasionally women's magazines mention surveying as a career (*Working Woman*, March 1985: 62–3, and May 1986: 20–1; and *She*, December 1987: 26). If women found out about surveying at school, it was often purely by chance. One woman surveyor, who went to a grammar school with separate boys' and girls' departments, said that she received a prospectus on surveying because it was sent to the wrong half of the school, being intended for a boy with the same surname. I have found frequently that those who entered 'by chance' (after school or later in their lives) were often more successful than those who had planned from the beginning to enter surveying, which is fascinating. The RICS commissioned a study of the image of surveyors held by career advisers and potential recruits in schools (Valin Pollen, 1986). Even the RICS admitted in response to the report that the general public were right in seeing the profession as 'middle-class, boring, and predominantly male' ('Surveyors stung by Pollen Survey', *Chartered Surveyor*, 29.5.86, Vol. 15, No. 9: 760).

Far from encouraging girls, many schools positively discouraged them from entering surveying. One woman said that she was made to feel naughty and disobedient for wanting to be a surveyor, indeed the teachers kept finding her leaflets about librarianship. Another said that the teachers sent a note home to her parents warning them that she wanted to be a surveyor and was therefore going through a rebellious phase, when in fact her father had first encouraged her. I have had several of these accounts, including ones from women who are still students, in spite of all the talk about getting more women into science and technology (Kelly, 1987). Many girls were warned by women teachers about lifting heavy equipment and going onto building sites. Women surveyors themselves have had to fight with schools to stop them pushing their daughters into doing cookery when they want them to do science. Many women surveyors expressed an extreme contempt for women schoolteachers, and in comparison saw surveying education as a form of liberation (echoes of Okely, 1978). It is easy to understand why many saw men, and not women, as the ones that had encouraged them, and therefore had little time for feminism.

Few colleges can resist showing the photogenic side of surveying in their prospectus. Typically, there will be a photograph of a mixed group of students out on site, wearing hard hats, with a female student looking through a theodolite. This creates two damaging false images, first, that surveying is macho-technological and to do with land survey, and second, that it is perfectly normal to find women in surveying. As stated, surveyors frequently use false images which exaggerate the numbers of women in surveying (and then believe this is the truth). A sensitising concept, in respect of this issue and the attitudes to women in general, is that of 'the abstract woman'. The men nowadays enthusiastically produce photographs of women surveyors (for example, as used in an RICS travelling exhibition on careers in surveying), and talk abstractly about the need to get more women into senior positions in surveying education and practice. But it all seems to relate to some hypothetical woman – a more 'perfect woman' than any existing woman surveyor; or to a place faraway where it is already happening. 'But, what about me?' ... 'don't be silly you couldn't possibly do that'. This 'abstract woman image' can be very destructive because it is almost as if women must 'believe' that she does exist somewhere, and they are being disloyal in feeling all is not fair.

More curiously, in everyday life (when men are not being asked to focus on women as a special topic) women will continue to be treated as non-gendered students. This may have benefits as in interviews for college they are unlikely to be asked about their future plans for combining two roles. However, this 'denial' of their existence as women in itself might be seen as off-putting. Many careers talks given by various colleges about a future life in the profession seem to relate entirely to a male pattern with no mention at all of opportunities for career breaks. Women are unlikely to raise such issues in the ensuing question time, as they feel awkward, or fear it will be used against them as a sign that they are not really serious. In fact some men lecturers do have prejudices about women's 'seriousness'. For example, one of the women in a highly technological area of surveying said that a male lecturer seemed to be extra harsh in the way he interviewed her. When she commented on this he apologised and said that he was only trying to put her off to see how serious she was (she already had good 'A' levels and relevant experience). It is unlikely that a male student would have been quizzed in the same way.

There have been a very few attempts amongst the landed professions to positively discriminate by setting up 'women only' courses, including one surveying course set up under the EEC Social Fund (*Chartered Surveyor*, 20.5.85, Vol. 11, No. 9: 674), which I visited. The students were mainly mature women, some of whom subsequently went into surveying, but others used it as a stepping-stone to other professional areas. More recently, another 'women only' course has been set up for heating and ventilating engineers, which is on the boundary with building services surveying. Whenever I have mentioned such courses to other women students they have been horrified, and say that studying (with) men is half the education.

Grades: are they different?

There may be 2,000 applicants for fifty places on the most popular courses (although many will be applying to other colleges at the same time; and at the other extreme, some courses have difficulty getting the minimum numbers necessary to run). Looking at their 'A' level subjects and grades, there seems to be little difference between the calibre of the applicants in respect of gender. However, it is noticeable that some colleges simply do not count art or domestic science at all on their points system. Many women seem to have got wise to what is expected of them and obtain the model set of 'A' levels including maths, science, and geography (echoes of Joseph's concept of 'precision'). However, these subjects may be combined with what may be seen as 'odd subjects', such as embroidery or drama, which may suggest to some men that such women 'are not really interested' or 'we must keep an eye on her'. Some women take additional quasi-professional subjects at school such as business studies, computing, and even 'O' level land surveying. This can be a trap, as taking such subjects can be seen as a sign of suitability for technician level. Likewise, some architectural courses do not count 'O' level technical drawing as a relevant examination.

The same subjects can be 'read' differently, depending on if they are taken by a boy or girl. I was surprised how many boys had biology and geography which are often seen as girls' subjects, but if boys do them they may be looked on in a different light as boys' subjects. Obviously the boys don't believe their own gender's propaganda about the need for science, that is for the girls' consumption. (These comments apply to the general practice courses, as on the specialist technological courses they must have relevant science subjects.)

What is amazing is the wide span of both subjects and grades possessed by students on surveying courses, although the average for estate management is around nine points (Whittington, 1987) but there are some exceptional students with three 'A' levels at Grade A, and a minority with three at Grade E on various surveying courses. It would seem that being the right type with the right interests and motivation is what matters more. If a girl (or boy) has low grades or inappropriate subjects and is still accepted, she must have some other plus factor, such as the right family background perhaps? Great emphasis is put on applicants' personal interests, especially sport (the key to career success and esteem, *Estates Times*, 24.7.87, No. 904: 14). This puts women in a rather difficult position, as one never knows how women's sports and interests will be interpreted. Some surveying courses like applicants that are doing the Duke of Edinburgh Gold Medal Award Scheme, and so the wise girl will do so.

Performance in college

The yearly examination grades and final degree classifications seemed to show very little difference in performance and distribution of marks between male and female students; if anything the girls did slightly better overall in some cases. (Interestingly, many course leaders commented that there was little correlation between good, or for that matter bad, 'A' level grades, and the marks gained in college.) Several of the course leaders that I contacted entered into the spirit of it and tried to carry out their own analysis and came to the same conclusion. As one male course leader put it, 'it would be unwise to suggest that these figures exhibit any trends whatsoever' and I had to agree. On the other hand, another male course leader claimed all the glory for increased numbers of women and their good grades. He seemed to imagine that the women got good grades because of his excellent influence, because they couldn't possibly have done it all on their little own.

The fact that the figures were the 'same' did not mean that women experienced surveying education in the same way as the males. Although occupying the same physical space, they do not share the same social space. You can end up with the same result through a variety of different factors. Also, the fact that women got good grades does not suggest that they liked it, 'my main aim was just to get through the course, as a means to an end, that's all'. There was also a very strong feeling from many women that men, particularly academically marginal ones, were 'always given the benefit of the doubt' whereas women were not always seen as 'serious students' and 'we have to work twice as hard to be seen to be any good'. Of course, academic ability can be used against a woman who is seen as being 'too clever', that is, not practical enough. I well remember being told that a certain student (a male as it happened) was 'too bookish', but nevertheless had potential to be a good student.

Double standards?

In the more technological areas where there are very few women, several course leaders have commented to me that 'the girls are either very good or very bad'. Because they have made a conscious decision to do an unusual subject for a woman and are highly motivated, and may overestimate what is expected of them, they work extra hard at the technological subjects and beat the boys. Needless to say, some of the boys are over-confident as to their knowledge and abilities in these areas and under-achieve. Also, subjects that are seen as difficult are often not really that complicated. However, some men were of the opinion that, 'sometimes women have trouble with strategic and analytical thinking, I think this is a function of their earlier secondary education'. Quite what is meant by 'analytical' needs considering: does it mean 'male' reasoning? Having talked to several of the women in question, and finding them capable of a highly sensitive sociological analysis of their lives, and comparing these with the somewhat mono-dimensional male students in the same groups, I am not convinced.

In conclusion, surveyors can allow a slack rein in education as they can compensate for this later. Education is generally seen as less of a problem than practice provided women let themselves be treated as non-gendered student units, and 'don't antagonise anybody'; 'but you've got to work twice as hard of course'. Several women commented that 'we were all students together, we got on as a team' but the very same women saw their treatment in the team in retrospect as less than equal: 'you accept anything when you are a student, it's only when you get out in practice and talk to others that you realise how different it might have been'.

Destination

Joseph observed little correlation between examination achievements and degree of seniority attained in the profession (1980). Several women have commented to me that their experience and achievements in education raised their expectations unrealistically, for they found that quite different virtues were valued in practice. Many were mystified as to why male colleagues, whom they judged as 'weak', were promoted and valued above them. Clearly the men were playing a different ball game with totally different rules. Educational success was not the name of the game. 'College is an unreal time zone, a suspension of reality.' Several men in education were aware of the problems that faced women in practice, 'the problems seem to be when they try to break through to partnership level'.

A final thought: some prestigious London practices take up to 30 per cent non-cognate graduates (with no prior knowledge of surveying), so the old argument used against women, about them not being adequately qualified or experienced, falls flat. As one very senior man said to me, 'as long as he is the right sort of chap, well, we can train him in all the little details, there's nothing to it'. I would be really interested to see whether in ten years' time some of these non-cognate 'chaps' had achieved partnership ahead of the cognate non-chaps, that is, the women that took surveying as their first degree.

Chapter seven

Fitting into surveying education

Introduction

This chapter looks at the place of women in surveying education within the territory of the subculture (the meso level on the model), and at women's personal experience of the situation (the micro level). It seeks to illustrate the interaction between the meso and micro levels, and shows, for example, how channelling and 'closure' mechanisms are already working at the undergraduate level, drawing on material related to a range of colleges via my three ethnographic approaches. These processes will, in turn, influence who ends up where within the different specialisms of surveying, and who is seen as suitable for what roles out in practice – and thus who has decision-making powers which in turn will shape the nature of the profession and the built environment (the macro level).

However, both women students and lecturers (including myself) are operating at two levels of reality. I found women were only too aware of 'the problems', and yet at another level they found aspects of surveying education, 'good . . . most rewarding' and 'thoroughly enjoyable and challenging intellectually'. Likewise, I actually like teaching surveying students and find the world of surveying education most interesting – but at other times encounter factors that are quite unexpected and unfathomable. I find both male and female students can be aware of social and environmental issues and are interesting and lively individuals; but at other times they seem on a completely different wavelength. As one man lecturer put it, 'all they want to do is get through the course and get their BMW!' Also, of course, different people's experience of the 'same' situation will differ considerably on the basis of gender, class, and individual characteristics. Some women seem to have more 'cultural capital' to draw on than others, in knowing how to cope and succeed.

Atmospheres and territories

The physical setting and atmosphere were frequently mentioned by women students, as these were the factors which they noticed when they first entered the territory of surveying education, and within which they subsequently operated each day. The women in one polytechnic surveying department commented that the 'brutalistic' nature of the architecture and the general environment could actually put women off (Sheffield, 1986) and this was reiterated elsewhere. On entering some surveying departments, particularly at the change of the hour when there are many students in the corridors, the appearance can be overwhelmingly 'male'. There is little 'space' for women as any common areas (especially those belonging to the students union) are likely to be male-commandeered, the territory sometimes being strewn with empty cups and cigarettes (echoes of the men's house, Ardener, 1981). Also, many women have pointed out to me, in diverse colleges, the inadequacy of women's lavatories as female numbers increase.

The regional location of a course lends it a particular nuance. People say that northern courses are 'tougher'; indeed, some still put considerable emphasis on traditional 'real surveying' (and have fewer women), whereas the south has moved more towards business- and management-oriented courses to produce the property professional that the City requires. However, neither a technological 'rough' ethos, nor a commercial 'smoothie' ethos, are likely to accommodate awareness of social or indeed spatial issues, let alone the subject of women.

The faculty and site in which surveying is located can affect the ethos of the course – surveying departments are found in a diverse range of academic locations. If surveying is in the same location as construction, it will create a very different climate than if it is located with law or business studies. This is not to say that one is less male than the other, but rather one manifestation of maleness might be more 'gentlemanly' and less 'macho' than the other, there being advantages and disadvantages in each for women. In one college, the women lecturers in town planning and surveying were convinced that when (thanks to their efforts) the numbers of women increased, they were purposely swamped by being put in with other courses with a much larger male majority, by rearranging the faculty structure. One can never prove this because there are so many other perfectly reasonable reasons why these changes might have been made, but one senses that something is wrong.

Surveying courses are all different, and there are all sorts of nuances and subuniverses to be taken into account, even between courses within the same department. For example, male quantity surveyors, housing managers, and general practice surveyors all dress slightly differently, but they are all surveyors. Women have a double problem here in 'precisely' interpreting how these predominantly 'male' differences should affect their own choice of appearance and demeanour. This 'matters' as they will be judged accordingly.

Some lecturers' 'manner' in dealing with students, and in meeting with each other can chill the atmosphere, a point raised by both women students and lecturers in a range of surveying specialisms. The desire to project an image of 'professionalism' is seen by many women students (and some women lecturers) as creating a sense of impersonality, regimentation, and officiousness, 'the coldness of the way they teach is a real barrier'. (This comment was not made only of male lecturers in some colleges.) 'I thought I knew him well, but in the meeting he addressed me as if he had never met me before.' Also, several women lecturers made the comment, 'male students never smile when you look at them like women do, there is no feedback'. (In contrast though, some male lecturers adopted a very 'chummy' one-of-the-boys approach, which was a mixed blessing for women.) Many women (mistakenly?) consider impersonality is the opposite of professionalism, which they imagine is about being 'good with people'. They find it difficult to square their expectations of a high value being put on individuality and the ability to 'paddle your own canoe', with the emphasis on 'team spirit' which pervades everything from student group projects to course management. Even if they go along with this, they may find that their contribution to the team has not 'counted', and they may be referred to as 'helpers' rather than seen as full members.

The normal way of college life

One of the problems with trying 'to make a fuss' is that many men surveyors genuinely believe that they are already very fair and progressive nowadays and 'trying their best'. Indeed, many welcome more women and want to encourage them, but may not know how to treat them when they do enrol. They would not see themselves as consciously perpetrating any anti-women conspiracy, but men 'behaving normally and treating everyone the same' is often seen by women as discriminatory, being based on male cultural norms. It was not so long ago that it was genuinely believed that 'letting more women in would lower the standard' (this was not just said by men, but also by certain 'patriarchal' women lecturers in several colleges), or 'it would make the course less professional' (for 'professional' read 'male'). Requests for anything else in addition to the right to be treated 'the same' are still seen as 'extras' or 'special'...'if women come on the course they must learn to fit in, they can't expect us to change everything just for them'. On the rare occasions that women staff or students have babies, such activities may be seen as 'bad planning on their part', 'inconvenient', or a sign that such women are unreliable and irresponsible. The same accusations can be heard against women out in practice, 'they never think it's the right time, whenever you do it someone will complain'. (Paradoxically, I have observed that women who already have children are actually seen as less threatening and 'more suitable' than those without.) Very few surveying or town planning departments have creche provision, and those that do are likely to achieve this by being on the same site as a more progressive department such as humanities. Whilst there are some moves in some colleges nowadays to improve the situation (not that anything much has actually materialised apart from talk) the scope of the whole issue is totally **underestimated**, and one hears comments and reports of comments such as 'it would only affect three mature students at the most' or 'the space is not available'.

Proposals for more part-time courses, which would be welcomed by many mature women students in particular, are often the subject of some misgivings or are seen as 'lowering the standard' (as, sometimes, are part-time women lecturers). Some men have excused the nature of the present full-time system by explaining to me that it grew up to meet the needs of the 18-year old school leaver, 'because that's the way it's always been, we didn't do it on purpose'. (Did I imagine all those years of part-time surveying education for people in offices, and the surveyor's natural antipathy towards full-time academic education, as described in the historical account?) However it's all 'horses for courses' as now the part-time route is being resurrected in some colleges, this often being aimed at mature students in order to maintain student intakes in view of the likely decline in the numbers of school leavers in the future. But as women have noted, 'the numbers will drop even more twenty years from now, if they continue to make it virtually impossible for women to have children and attend college or have a professional job at the same time'.

In contrast, men's lifestyle is accepted and catered for as part of normal life. If certain male students want to go and play rugby or golf, and leave early of a Wednesday afternoon (which is not really allowed), they will simply come and tell me so, 'Mrs Greed, I have to leave at 11.30, as I have got to go and play golf' (as one man told me, there are only really two types of surveyor: those that play rugby and those that play golf). If I object they will looked puzzled, even offended, and make some comment like 'but it's sport' as if that gives them a divine right to leave early and I am preventing them from doing their duty. In complete contrast I have had instances of women students with children (few though they be) coming to me and stating with some embarrassment, 'I hope you won't mind, but you see my childminder is ill and well there's nothing else I can do, if I could be allowed to leave early'. They seem to assume that women in authority are not necessarily on the side of women.

Likewise, rugby injuries are looked upon with sympathy, even admiration, whereas women's illnesses are often looked on with suspicion or as a sign that women are not tough enough. I remember a student group asking me to let them 'recover and warm up first' when my lecture followed a land survey session in mid-winter. When another group didn't go out, they told me, 'the lecturers thought the girls shouldn't go out in this weather'. The boys had shifted their own 'weakness' onto the girls, and then saw them as inferior to themselves. In comparison, I remember when I was working as a planner that I returned frozen and worn out after being out on site. I told my team leader, that I had 'to recover first'. He wanted me to do something else straightaway without even a coffee, 'but you haven't uncovered yet', he said refusing my request with characteristic innuendo.

The women who fit into this sporting subculture need to be sporty types themselves, as one male student put it, 'professionals are more athletic'. However, women's sports can be seen as a joke. If the girls join in unisex games this might also be to their disadvantage as boys have been heard to say, 'oh girls are good at that, that might be a handicap'. It would seem better to choose a high status unisex sport such as sailing, wind-surfing, or hang-gliding that one is **not** too good at. One woman, now in practice, explained the importance of this to me saying, 'we all went through a dangerous sports phase, but now the most I do is help to organise the squash ladder'. The important thing is to be involved and show some interest in sport. Some women are mystified as to why most colleges allocate a whole afternoon per week for sports, when they may have been told at school that an enthusiasm for sports is a sign of being unacademic, and to spend time on yourself is a sign of selfishness. Related to this there is a strong emphasis on the 'pub culture' (Hey, 1986) which is seen as 'normal' (and essential to forming lifelong male friendships) but which many women, particularly older ones, have little time for (especially if they want to get through the course as quickly as possible). Indeed, 'socialising' is a very important aspect of surveying but women often underestimate or despise it. Whilst some of the women have said that they see the boys in their year as 'stupid and immature', they believe that the boys and many of the lecturers see them as being of little consequence, 'rather childish and sweet'. Quiet studious girls are often seen as not very interested (or interesting?), whereas noisy and easy-going boys are seen as good chaps.

The women who do well are those that learn the arts of survival and camouflage 'to blend in'. Many adopt the unisex uniform of jeans and even the rare older mature women students appear to become 'younger' or at least blend into the youth culture (not necessarily a bad thing?). Women are still judged more than men on the basis of what they wear and some men lecturers consider themselves entitled to say so, 'she won't be able to dress like that when she goes into practice', but at the same time seem to be much more lenient towards boys who are seen 'as going through a phase'. A theme that came to me again and again was that although women in surveying are not 'feminist', they **do know** the situation is unjust; but they know how to play the game. Rather than challenging the system or complaining, they learn the best strategies to survive, negotiate, 'manage' problems, and succeed. They may therefore be unwilling to talk about their difficulties to other female students or lecturers, particularly not on male territory within the department. This can be confusing for other women who have been used to more open sharing of problems, and therefore conclude everyone else is getting on OK except them.

When I look at the department running like clockwork and the students of both genders busily working away in groups, and drinking coffee together, I sometimes feel quite neurotic about imagining that anything might be wrong. The students seem to work together as a dynamic machine when project work is in full swing, both male and female students showing great enthusiasm for commercial property development projects which emulate the situation in the 'real world'. Perhaps they are right, it is only me! But there is tension there. I tracked down some of the women who had dropped out of surveying education, as I suspected their experiences might be more valuable in understanding what is wrong. The general opinion was 'it's not overt sexism and discrimination, it's all the little things, you can't put your finger on it'. Several were vaguer in describing a general feeling of 'not really knowing what it was all about and where I was going'. These were not casual students, but women who were orginally very determined to do surveying. I suspect some of them experienced 'culture shock'. Several described how 'it was all so different from anything I'd ever experienced. I don't think I ever got over it'. This was particularly true of women without brothers and 'men' in their background. There were also a few instances of actual harassment in various colleges. Women were undoubtedly aware of 'what might happen' but 'we don't like to emphasise it, as it might draw attention to it'. Indeed, more radical women on housing and town planning courses were seen as letting the side down, in drawing attention to it. Perhaps if no one mentioned that women were 'different' no one would notice – 'heads under parapets' (SBP, 1987).

The role of women

The position of women in surveying education prefigures the situation in practice. Women students and lecturers have to carefully avoid being categorised as 'helpers' (the handmaiden, 'helpmeet', Heine, 1987) whilst at the same time they must be seen as 'pulling their weight', although even then they may find that, when it comes to the crunch, their efforts 'don't count'...'oh you haven't done anything'. (Umpteen women surveyors gave their own versions of this one.) In both education and practice, women can get stuck in accepting this helpmeet role; but paradoxically if they are really good at it and are not seen as threatening, odd, or feminist, they might be promoted to a higher level of it. In contrast, other 'exceptional women' do actually break out of this and exercise authority in their own right.

There is a strong emphasis on group work, especially practical group project work in surveying education. Within project groups, all students are doing 'equal' work to achieve a group mark. Women may be given the secretary role in a group; or if there is anything 'social' or sociological in the project they may be allocated that task. The men might be initially friendly, but gradually the women find themselves doing everything (friendly and apparently non-sexist behaviour can prove the most patriarchal). A woman student may find herself being given all the responsibility and all the blame for organising everyone else, 'it's just like at home, my brothers expect me to pick up their socks for them and here they wouldn't do any work on the project until I had sorted it out for them'. More cunningly, some teams will make the woman the leader rather than the secretary, but then she may find herself doing even more work, whereas a male leader may do very little. I drew a comparison with an example given to me by a woman I met at a conference, who ran outward bound courses for young people (from deprived backgrounds). In spite of women always being told that they lack visual spatial abilities (echoes of Maccoby, 1972), both in the case of her students and mine (in spite of class and educational differences), a girl was given the job of map-reader and the boys would wait for her to direct them. Needless to say, in the final project presentation of the work the men did most of the talking. I even had a case of a student telling me he could not do his talk properly because, 'she has not had enough time to prepare my notes'.

One cannot generalise as sometimes it can all go the other way and the male students do more than their fair share throughout. What is fascinating is that as numbers of women increase, some women seem to get lumbered with traditional secretary roles, whilst, curiously, others may be treated the 'same' as the boys. There are many personality factors that need to be taken into account above the baseline of gender. Listening to my students over the last few years, I have observed a great range of variations of behaviour. Sometimes I wonder if it is a trap to attribute importance to manifested behaviour, as women in practice have commented that 'things happen, completely unrelated to how people behave, your face fits or it doesn't, there are no rules, they are just playing with us'.

It is interesting that when there is a free choice of members for project groups 'women choose their friends, whereas the men choose whoever they think is best whether they like them or not'. A lone woman or overseas student might get snapped up instantly if they are seen to be 'good', otherwise they might be left altogether. If one is a black male and good at sport, all else may be forgiven.

Some projects have to be presented as written reports and an impersonal style is expected. Many women have been ill-equipped for this, being taught in girls' schools to write 'essays' for English in a more literary and personal style. Women have to learn how to change this and do a million other strange things, and 'quickly', but no extra praise is given for this achievement. Some women students who pride themselves on neat handwriting and attractive appearance are surprised when this is seen as a sign of lack of seriousness (although paradoxically attractiveness can be an advantage out in practice). Male messiness may be seen as a sign that the males are too serious to be tidy (unless it's really bad). One cannot generalise too much as some boys are far 'tidier' than girls, and some are academically far better than the girls too, whilst some of the most 'agricultural hand' types of handwriting belong to women.

There is a strong tradition in surveying, architecture, and town planning courses to mark coursework by means of a 'crit session' in which students must defend their work whilst it is being verbally torn apart by lecturers and rival teams. This can be an unexpected and unsettling experience for women students used to individual work and a less adversarial approach to gaining marks. I must admit I was quite 'good' at this, and was 'equally' aggressive to both male and female students before realising how some women students were taking it. Indeed I, and other women lecturers with whom I have spoken, did not realise, at first, that many men too could not cope with 'critical women'. Women lecturers behaving the 'same' as men can be received negatively by both male and female students, who might take it more personally than from a man, 'who does she think she is?' This could potentially turn them against town planning itself, with which surveyors have a love/hate relationship to start with. Is this another example of women unwittingly fulfilling the requirements of the patriarchal subculture? Several women in practice have also told me that they must adopt a totally different demeanour with family members because an assertive answer might be seen as 'rudeness' or 'sharpness' outside of the professional arena.

Women students often get given the more 'social' component of an assignment. Likewise, whatever subject a woman teaches, even the more technological ones, they may be called upon to do more social type teaching such as communications projects, seminars, liberal studies, and social aspects. Women lecturers seem to be considered more naturally literate than men and will be expected to be good at 'English', and maybe first aid too. Women who teach more technological subjects may find their work being questioned more frequently, or even students 'checking' with male lecturers that the woman's lectures are 'right'.

Another sensitising concept that indicates the position of women, which I came across frequently, is the use of 'we'. One woman lecturer, who had been at a college several years and had done a great deal of work in building up her department, was surprised to be told by a male colleague, 'oh, we sometimes let women lecturers in', as if she had not been there before, and was not part of 'we'. More amazingly, when I myself asked a new little chap on the first year in the first week of term why he chose surveying (to be friendly), he replied, totally out of context, 'oh, we sometimes allow married women to work', no doubt feeling threatened by the existence of 'too many' women lecturers. A variation of this is for boys, who have perhaps spent two weeks in their uncle's office over the summer before coming on the course, to state, 'in practice we do such and such', implying that we know nothing of practice and they do.

As both women students and lecturers have explained to me, at the beginning of the course (which usually runs for three years) the students have to establish a pecking order between themselves. One way of achieving this is to put down women lecturers and other people that are 'different'. After some 'readjustment' on all sides, women are likely to be treated 'normally' by the male students by the end of the first year, until the third year. Then the promise of higher status in the real world presses in upon them, and some male students will 'change' again to become quite 'superior' to both women students and lecturers, almost as if they are seeking a new identity, and they construct this by going back to looking down on women in the department, 'I'm sure he blew smoke in my face on purpose'. Of course, if a female lecturer is responsible for supervising their final year dissertation, they cannot go too far, for, paradoxically, these sexist attitudes are actually controlled by the patriarchal organisational structures within which education operates.

Classroom interaction

I tried to keep track of which students spoke, how often, and when. Sometimes I got so interested in what they were saying that I completely forgot. But it did seem that overall men spoke more and definitely interrupted more. Even when the numbers of women in groups increased, they seemed to be less visible or possibly attended less, whereas in the past the one exceptional woman was much more visible. A rather quiet first-year female student was slowly struggling her way through a paper on regional planning in a tutorial, when, suddenly, one of the more assertive males looked directly at me, engaging my attention with great urgency (thinks: the building is on fire and I haven't noticed): 'Mrs Greed, my neighbour has just received an enforcement notice about his panel-beating business, the planners say that it counts as a change of use, what should he do?' I was completely taken aback, and like an obedient little teacher my mind was already going through relevant planning law. But stop: there is a totally bewildered female student looking quite exasperated, plus a disturbed group some of whom are full of admiration for the culprit, 'discuss it with me afterwards' – he looks hurt – 'I do like people to relate planning to real examples'; he looks happier – 'but we were doing something else in this tutorial'. Such is the social conditioning that it's impossible not to let a male student totally destroy a tutorial (in the most gentlemanly way possible).

Some women seem to land themselves in it unintentionally (out of nerves?), such as the one who gave a talk on 'improperly mixed concrete', or they will make self-derogatory comments, 'I'll do it in my own dim little way'. By this, the woman seems to be demonstrating that she is not to be seen as a challenge. Some women are giggled at when they speak, whereas others are taken much more seriously and treated 'the same'. I am still pursuing the reasons for this but it would seem that some women are perceived as silly **because** the men are silly towards them, and they have been picked on, because the men find them potentially threatening or unusual, that is, very unsilly. On the other hand some men who seem 'a bit wet' and go around with the women might also be laughed at. Also one sometimes gets an older or more mature woman who virtually takes over the group, and raises the standard. The men seem to be cowed by this and accept it sullenly, almost as if they know it is good for them (that is, it is not really a position of power but an inverted helpmeet role; like a mother). However, I have observed that such women might not do as well out in practice as those they dominated and thus helped.

Likewise, in 'grown-up' staff meetings and day-to-day interaction between lecturers, many of the same attitudes and occurrences can be observed. Several women lecturers have commented that a man and a woman staff member may be asked to do the same activity, such as that of being a year tutor. A woman may feel she has been saddled with a dead-end counselling role, but for a man the same job may be the first step on the vertical ladder up to course leadership, 'you can't win, if you do anything they will always see it differently from if a man does the same thing'. One can be independent and initiate something and still find you are seen as the 'helper' in your own creation, a common complaint. In meetings 'you say something really good and they ignore it and carry on speaking as if they had not heard you. Then ten minutes later a man says exactly the same thing and everyone stops and says it's a wonderful idea'. Also, several women have commented that in the rare meetings that women chair, the men may concede to use 'chairperson', but then continue as if women are not part of the context which they are discussing either as students, staff, or people in society, 'it means nothing'. I had many accounts of such instances from women in practice too. All the elements of not being heard, and being rendered invisible, being 'wound up', and not being taken seriously, as described in the feminist literature (Spender and Spender, 1983; Cline and Spender, 1987) are clearly evident. This must ultimately influence professional practice, 'if they won't hear you or see you, they aren't going to listen to what you have to say'. Nevertheless, some women overcome these problems and succeed.

The subject of women

Women may become very embarrassed in class if 'women' are discussed as an issue. They will either seek to disassociate themselves from 'those sorts of women' (who presumably live on council estates) or go completely silent. Once when I suggested 'women and planning' as a seminar issue, one girl in the group let out a shriek of embarrassment. If one seeks to discuss the 'ethnic' issue and the inner city, I have often noted that some black overseas students will totally distance themselves from 'those sort of people' and look very embarrassed. A similar situation may arise if there is someone of the 'wrong' class in the group, 'I expect you live in one of those sweet little terraced houses'. There are, however, occasions when 'minority' students speak their mind. One example, which combines both race, class, and gender, was as follows:

There was once a most self-confident, socially aware, black female student in an otherwise totally white and predominantly male group. She started giving some very insightful comments in a tutorial. Suddenly, one of the more aggressive public school white males, not previously known for his social awareness, cut right across her. Addressing me directly and loudly, he declared how important race was and how black people should be asked for their views more often. He looked at me as if he wanted me to praise him, as if he was jealous that I had given my attention to the woman whom he had interrupted. The black student was left upstaged, and it seemed that the whole class had rallied to the side of the white student. The desire of some white middle-class males to take control of, and 'solve', the problems of others creates a sort of inverted colonialism (the spirit of the colonial land surveyor lives again). As some more radical women students have put it, 'they don't care a damn about women or black people, but they've got to feel that they are in charge all the time'. The desire for control, either through the market, or through social policy, seems to be a factor that unites certain types of surveyors from both ends of the spectrum. It explains why people who superficially appear to be ideologically incompatible can form powerful alliances to their mutual interest in the maintenance of patriarchy and 'more jobs for the boys' (two sides of the same coin – another link for the model). Other men surveyors who are not interested in these power games simply want to do their professional duty and get on with their work 'like women'.

Women students can say and write the oddest comments about 'women', almost as if they see themselves as the typical male practitioner (and not as women) in their mind's eye. Perhaps part of the professional persona is not to personalise matters, but to act as a disembodied authority with no needs or wants of one's own. Women lecturers may encourage this, one had written on a female student's essay, 'try to play down your own views, the examiners may not like them'. Also, many surveyors talk about 'females' rather than 'women', when discussing both professional data and their personal lives, which is also a symptom of the impersonal ethos.

I get the impression that neither the male, nor, surprisingly, the female, students associate what is said about 'women' in the abstract with their classmates. I remember a woman describing an area as 'a bad area because the women go out to work' when she intended to do so. Many women noted that there seems to be an inbuilt assumption that 'women', and most especially housewives, are 'a problem' along with 'blacks' and 'the unemployed'.

In comparison, I can never forget how, when I was a planning student, just about every lecture (whatever the subject) seemed to include comments to the effect that all the world's problems were women's fault for causing over-population (thus legitimating the need for rational men to take control through planning – and sowing the seeds of the present demographic crisis). It was constantly said that 'women have too many children', this often being prefaced by the inference that men were blameless because women seduced men and asked for it. The fact that it was all women's fault removed the need to do anything for women because 'they bring it on themselves'; therefore emerging feminist town planning policies (and planners) were instantly verbally massacred in this climate. Surprisingly, it was suggested that nurseries and child care should be provided, not for the benefit of women, but rather because this was more efficient to enable women 'to work' and because women were such a bad influence on children. In particular, the comments 'housewives are stupid' and 'women are selfish to have children' seemed to be repeated endlessly. Indeed, it would seem that some of the lecturers wanted to do without women altogether. They would wax lyrical about the future possibility of babies grown in testtubes (which apparently did not cause over-population). Nowadays I still hear reports from women students of such attitudes being perpetrated in architecture and town planning courses. In comparison, surveying is light relief. As surveying is 'peopleless' and more 'gentlemanly', and has less of a social and policy content, there is less opportunity for sexist comments and anti-women policies to pervade the discourse.

I noted the comments my students made in my planning tutorials as to their perceptions of women's needs. The very question of provision for women is often seen, in itself, as a sign that women are selfish, and middle class. Nurseries or child care facilities on housing estates are seen as indulgently middle class, even if intended for working-class families; but of course pubs and sports facilities are seen as normal and essential.

One chap went on and on in a tutorial about every conceivable problem related to women living in flats. I was watching the female students, and they were silently giggling at the silly things he was saying. When I questioned him as to why there were no men in the block and who these strange women were, he replied, 'Oh, I did women, because I thought that's what you wanted, because, well, they're all women in this group' (a gross exaggeration). On the other hand, according to women in various colleges, a minority of boys will try and shock women lecturers and students by scattering references to prostitution and rape (and even child molestation) through their tutorial discussion. This can seem almost amusing to the women as it appears that the boys think that the girls share the same fear and unfamiliarity with 'the female' as they themselves. Nowadays, some town planning and housing courses include something on 'the changing role of women' as taught by men (the RTPI now requires that the needs of 'minorities' are paid lip-service in planning education). This is seen as enlightened by some men but can be quite soul-destroying for the women listening if the impression is given that women now have more opportunity to succeed and improve themselves because 'good men' are willing to help them. Some women find the 'hopeless efforts' of such men 'quite hilarious', but would never tell them so to their face; indeed, they might appear to be impressed so as not to antagonise them.

Other people

I was interested in surveyors' world view and attitudes to others because, as stated earlier, this would affect their perceptions in professional decision-making of what urban society and the built environment 'needed'. There are some strange attitudes manifested by students regarding social class. Some students see themselves as quite 'ordinary' and assume anyone that does not succeed or is unemployed 'has not tried hard enough at school'. Housewives and the unemployed are often lumped together as 'the less active members of society'. Some seem to come from very sheltered backgrounds, 'I know there is no unemployment in Lancashire, because I live there'. There are many other inbuilt assumptions that come out from time to time, such as 'everyone has a car nowadays', 'houses are not expensive', or 'everyone wants to play football', which will all in due course feed their way into their approach to professional decision-making when they go out in practice. There is, however, a willingness to learn and a strong social awareness among some of the students. Some of the male students even seem far more aware and open in their views than some of the women.

Sometimes students appear to show a social awareness by raising perfectly valid issues and then quite innocently propose an inappropriate solution. One student correctly noted that there were no local shops or amenities provided near to a rather run-down block of council flats. He identified the social problem as follows, 'therefore the housewife will have to drive further to the shops' (in an area where very few people owned cars).

When students do show social awareness, it is likely to be framed within the viewpoint of traditional male social policy; 'sport', in particular is seen as the panacea for all social problems as if the world were composed entirely of young males (like themselves but of a different class). Likewise, students seem to 'believe' urban sociological literature that stresses the importance of working-class communities, 'mum', and 'work', which of course only give a partial, 'malestream', and arguably conservative world view.

It is highly significant that town planning has been separated from surveying into two sets of courses and two sets of professions. Surveyors are alleviated of the necessity to be concerned with the non-profit-making aspects of the social and spatial structure of cities because, as several women and men surveyors have commented, wide-eyed and innocent, 'that's the job of the planners isn't it?' This attitude trickles down to education; indeed, many students come on the course already convinced that planning is bad, unnecessary, or irrelevant, 'there is a great deal of consumer resistance' as one man lecturer put it. However, strangely, the students may know about and support some aspects of town planning, but 'that's because they think it's geography', which is a popular school subject with students. In fact one of the roles of first-year planning is to help the students sort out for themselves what is the scope of planning, geography, and surveying, 'so they don't support the wrong team', or score an own goal.

Subjects taught

Most general practice surveying courses will include the following subjects: valuations, urban economics, property law, town planning and development, construction and services, estate management, computing, a little land survey, and in some colleges an element of management and business studies. (Note that my subject, planning, therefore makes up but one part of the whole and is one of the few subjects with a social component.) It is noticeable that students do not interconnect the subjects: they are fragmented in their minds. To take a non-gendered example, they may not realise that the rights under private property law must be seen within the context of national town planning law. More worrying, they may not see the implications of what they learnt from my 'social aspects of planning' for property development projects later in the course, 'we did that last year so we didn't think it was relevant'. Likewise, many 'grown-up' surveyors believe, 'social issues are nothing to do with the market'.

When looking at urban development, the impression was often given by both students and lecturers that some impersonal force had made it happen: not human developers who were meeting the needs of other human beings for whom development is. The abstraction and 'peoplelessness' of it were observable themes. If students had studied human geography at school, it created the right 'peopleless' attitude that could be built on at college. (The people in geography seemed to be subsumed under areas, classes, or problems, and never seemed to surface as real people.) This 'peoplelessness' softens them up for accepting another 'peopleless' world view, namely, a commercial professional one. Although they can transfer the 'impersonal' element easily from geography to surveying in the first year, it can take until the next year before they think from a commercial market viewpoint rather than a geographical perspective. Indeed, many surveying lecturers wish that school geography was more commercially-oriented for this reason.

All this seemed less true of the women students. If, at the beginning of the course, I ask male students what are the social issues in planning, they will inevitably refer to abstract issues such as regional planning. Females are much more likely to refer to the need for community facilities (which actually confirms gender stereotypes). However, give it a term or even a year, and women in particular will start purposely choosing impersonal approaches to topics. I have had several incidences of women discussing problems about residential estates purely in terms of building and maintenance issues, when they earlier showed concern with how people lived in the development. At the same time, some of the impersonally-oriented boys are 'learning to play the game' and integrating social and commercial issues into their work (so the girls are still one step behind, even after swapping rides).

Other subjects bear no relationship to what they did in school. Valuations, although a totally impersonal subject, creates a whole new way of looking at life based on seeing everything in terms of monetary value. It 'appears' to be a mathematical subject and can therefore be used as an exclusionary mechanism in itself. Some valuers say it is not really as difficult as it is made out to be. Nowadays there are also computer programmes that take the hassle out of valuation. Indeed, some valuers say the most important skill is the one of exercising professional judgement based on an awareness of what is happening in the world of property. (At its most basic this may involve less experienced valuers using the unofficial 'comparative method' of phoning up a colleague and finding out what the going rate is!) In fact some of the boys are quite poor at maths, much to the annoyance of some of the women students in the same project groups, 'they don't realise, they think they're wonderful, as they've never met anybody better'. One chap said he wanted to be a mathematician but had chosen surveying because he thought 'mathematicians aren't professionals'.

Surveying students like factual subjects in which there is one right answer and a fixed area to learn. Like Joseph, I had imagined that planning would be seen as one of the most 'waffly' subjects, and 'hard' subjects such as construction would be seen as the most straightforward. On the contrary! It would seem that much town planning has hardened up to reflect the greater emphasis on partnership between the public and private sector within the enterprise culture. Amusingly, several women ex-students from a distant college told me that construction was the most waffly subject, and that the man that taught it was seen by them as a 'complete wally'. (One was given the distinct impression that he saw the girls as 'weak students' and himself as 'a real man'.)

The fact that I have many reports of all sorts of sexist comments by construction and technology lecturers up and down the land is not surprising (but paradoxically a minority of individual male construction lecturers tried extra hard to be aware of these issues). What is surprising is that the women take it all in their stride and see it as an occupational hazard to joke about, 'now girls, you will know about hot water systems because you know about airing cupboards don't you', 'I won't ask you this question as you don't even know how to change a plug'. Some women answer back. One lecturer told a female student that 'mixing concrete is like mixing a cake' to which she replied, 'I didn't know that you baked cakes'. Another told a woman that she lacked visual spatial ability (Maccoby, 1972). She told me, ' I told him, "don't be silly, my sister's an architect and she is doing very well, much better than you". That shut him up'. You have to be brave and foolhardy to do this; indeed, most women prefer to ignore it. When similar things happen in practice they are already conditioned to accept it. On the other hand, I have several examples of boys doing the daftest things such as putting outside windows in the wall of an inside corridor on a plan in their project work. I'm sure if a girl had done that, we would have never heard the last of it.

Just occasionally, one comes across examples of 'real people' being used to liven up lectures. For example, stereotypes were being used by lecturers, such as 'Mildred the militant housewife' who married too early and therefore blamed society for her own mistakes. There was also an example used in a lecture of a 50-year-old woman senior partner with three children, whom I very much doubt exists, but was given as a typical example (remember only about 600 women surveyors are over 30; I estimate about half of them have no children and the majority of the rest have one or two). Such examples are most likely to crop up in management studies where there is a people element of sorts. More positively, in a valuation examination question another lecturer used the example of a small surveying firm in which one of the three partners was a woman, giving a very sensible role model.

Subjects that are associated with high status professions outside surveying or are associated with high status property, land, and wealth are popular. Those aspects that one would imagine students would find irrelevant because they are ancient and of limited practical use today, except to a very small percentage of the population, such as the laws relating to settled land and hereditary estates (which affect perhaps 50,000 people at the most) are welcomed, because they reinforce subcultural values. My subject which affects over 50,000,000 people, is seen as definitely less relevant because it is 'social', and 'less factual' and therefore not as real. It is often the women students who like law best, and strangely they seem to identify with the principal male actors and do not even consider the very obvious (but seldom stated) fact that women effectively had no right to own anything for many centuries. Some women confirmed gender stereotypes by declaring to me that it was the most interesting subject because 'I could get on with it on my own'. Some women students get very upset about law lecturers 'livening up lectures' by using sexist examples, but seem to miss deeper structural issues related to patriarchal law itself.

Conclusion

It would seem that the men genuinely want to increase the numbers of women on courses and within surveying itself, but when they get them many do not know how to treat them. They either 'see' them as 'the same' as the male students, ignoring both their 'special needs' and the way they are likely to react to the 'maleness' of lecture material and service delivery, or they treat them as 'different' on the basis of outmoded *gallante* attitudes or inaccurate stereotypes which limit their role and development within the educational setting.

Chapter eight

The position of women in surveying practice

Introduction

As can be seen from Table I, the bulk of surveyors are located in the general practice division (GP), followed by the quantity surveying division (QS), and then by a series of smaller divisions, including planning and development (P&D). (Housing counts as an 'option' with the GP division.) Note that there is a disproportionately high percentage of women in general practice.

Table 1 Percentage of surveyors by division, 1986

Category	JO membership		All membership	
	Female	Male	Female	Male
Building surveyors	5.6	10.0	6.2	8.6
General practice	76.9	47.0	65.0	41.5
Land agency	3.2	8.5	3.6	5.8
Land surveyors	0.8	0.5	1.2	1.9
Planning and development	4.4	1.0	2.9	2.1
Quantity surveyors	9.1	33.0	20.9	38.7
Minerals	<.1	0.5	<.1	1.0
Total	100.0	100.0	100.0	100.0

Explanation: For example, 6.2 per cent of all female surveyors are in building surveying, but as shown in Appendix 1, only around 3 per cent of all building surveyors are female.
Source: (JO, RICS, 1986: 11; based on data, received with thanks, from the RICS Record Office.) These proportions have not changed significantly over the last three years, and comparable figures were most readily available for 1986 between the JO (members under 33 years of age) and the RICS as a whole. For total RICS membership percentages for 1989 see Appendix 1.

Education and practice today

It is a false impression, as one prominent woman surveyor imagined, that because only 2.9 per cent of all women surveyors are in the planning and development division (P&D), the position of women in surveying therefore has very little effect on what is built. The whole ethos of the surveying profession and many of the professional decisions in all the divisions affect women as consumers of the built environment. In particular, the commercial 'perspective' which predominates in both the general practice and quantity surveying divisions may overshadow a wider outlook, the decisions as to the viability of many land use and development proposals being based, of necessity, on financial considerations alone. Also, the separate P&D division was created less than ten years ago and has received a mixed reception, and indeed many surveyors who have stayed in the general practice division are in fact doing P&D work.

The Junior Organisation has noted that those women who are in P&D are likely, on average, to receive higher salaries than the rest as many are employed by investment and finance companies (JO, RICS, 1986: 5, 1988: 29) which are seeking development 'opportunities' rather than to meet social or spatial need. The sort of women working in this area are likely to be either rather 'up-market' and involved in, for example, dockland 'yuppification', or alternatively, bright 'back room boys', doing research on lower salaries, or in local government.

The modern surveyor

What exactly surveyors 'do' (in any of these divisions) is a bit of a mystery (even to surveyors); basically if a surveyor is doing it, that is what surveyors do. Much of what they do is scarcely related to 'land' and might be equally undertaken by an accountant, lawyer, or general manager. Surveyors nowadays see themselves as the new property professionals and investment accountants of post Big Bang optimism (Avis and Gibson, 1987; *Estates Times*, 24.10.86, No. 867: 14–15), a theme reflected in many articles over the last few years (e.g. 'Surveying the new frontier', *Chartered Surveyor*, 19.6.86, Vol. 15, No. 12: 976; 'The signs of an end of an era', *Estates Times*, 7.11.86, No. 869: 17) and discussions about 'the opposition' ('The Big Bang', *Chartered Surveyor*, 12.2.87, Vol. 18, No. 6: 63). Much of the old-time technical toil has been removed with the introduction of computers and shifting of much routine work to technician grades, to enable the surveyor to concentrate on professional issues. Indeed, it was felt that, 'they've got to make it more sexy and exciting if they don't want to end up as technicians' (*Estates Times*, 25.7.86, No. 855: 8). As the profession has progressed, there is more 'space' for women. Women who were put off by the rather macho-technological image of the past are now much more attracted by the smoothie office image of the property professional, and do not seem intimidated by the use of computers. 'They have taken the wellies out of surveying' as a male colleague commented to me, 'no more wellying about on muddy building sites' (but wellies are still cult objects, e.g. 'Winging a winning wellie', *Estates Times*, 22.8.86, No. 859: 7).

The nature of surveying firms

Surveying firms come in two main types. First, there are the large prestigious practices, mainly located in London, which may have as many as fifty to a hundred full partners (with an inner sanctum of very senior partners) and a vast pool of associate partners below that, and then below that again numerous young surveyors. In the enterprise eighties, the numbers of new recruits taken on by some large practices, in any one year, can be as great as a number equivalent to a third of their total existing 'man'power establishment. This is exceptional and far less common among smaller firms. Several of the large practices are now going over to a corporate structure which many women welcome as they believe that promotion will be more likely to be based on formalised criteria than on grace and favour. Some surveyors believe that in the future the majority of practices will go over to a corporate structure (no doubt with men mainly in control). The large London firms deal primarily with large corporate clients and landowners, and are unlikely to undertake any residential estate agency. Some large firms combine their general practice commercial activities with building and quantity surveying, but in most cases this is not so and there are other specialist prestigious London practices covering these areas. Again, these will deal with 'big projects' and they would not be seen dead doing a normal house survey. Most of the large firms, whatever their specialism, are primarily dealing with 'good commercial property', that is, prestige office, shop, and industrial developments. As with people, buildings come in different social classes!

Second, there are literally thousands of smaller firms throughout the country, as will be seen by glancing through the RICS Year Book, ranging from sole practices to provincial firms with perhaps three to twenty partners. There are a few prestigious provincial practices which are much larger, and often 'ancient', and in the same elite league as the best London practices. But not all practices in London are large, and there are hundreds of smaller practices in London and the Home Counties too. Many of these, both in the London area and the provinces (not all), have a strong estate agency component, and it is these that are being gobbled up by banks, building societies, legal firms, and other financial institutions since the deregulation of financial services (*Estates Times*,

12.2.88, No. 931: 10). A few large national monopolies are developing. Some predict that we will end up with a situation similar to that of the United States of America, where estate agencies (realtors) are dominated by one large company which operates on a franchise basis. Women have mixed feelings about this, although ironically many of the franchise-holders are women. In Britain, some believe that the influence of the banks and building societies, which are traditionally employers of large female workforces (albeit non-professional), may be to their advantage. Others feel it makes it even more difficult for women to set up their own independent businesses. As one woman owner of an estate agency business told me (who had been pressurised to sell out to one of these chains) 'all they wanted was the site of the business, and all my staff under the age of thirty . . . they didn't want me for a start!' Many women try their best to avoid 'resy', that is, residential work of any kind as it is seen as low status and quasi-professional, but there is no denying it does provide a viable career alternative for many women surveyors who have 'dropped out of the London rat race' (usually because of home commitments or husbands that move around the country pursuing their careers). The current trend to hive off estate agency as a non-professional department quite separate from mainstream surveying, 'so that anybody can do it' may be to the detriment of the professional woman who has established her own 'niche' in this area. Of course, all estate agencies are vulnerable to the ups and downs of the residential market, especially changes in interest rates and house prices; it can be a very risky business venture for any woman or man.

Distribution of women surveyors

As can be seen from Table 2, over a third of all surveyors are located in London and the south-east. I could not get figures for all regions for women, but, in comparison, the JO notes that 42 per cent of women surveyors are located in London and the Home Counties (JO, RICS, 1986: 11). (The majority of our women students now go to London.)

Overall, it would appear that there is a greater predominance of women in the south as a whole (than for the men) and far fewer women in the north, 'there's hardly anybody between Manchester and Edinburgh'. Overall figures give a false impression, as women will be dotted about different firms within the region rather than grouped together. In some firms, especially in London, 15 per cent of their young surveyors will be women, but others will have far fewer women ('the typists don't count of course'). But some say, 'we prefer it like that, I don't like being with a gaggle of women, I prefer working with men'. For those that would like to work with, or at least know of, other women in their area, it can create a sense of isolation and fragmentation. In some instances, women from different sections of the same firm met for the first time as a result of my arranging a meeting with them all. Within provincial firms there are far fewer women, except for a few concentrations in the more prestigious firms in the larger cities, 'you can count them on the fingers of one hand'. In these, 3–10 per cent of surveyors will be women which may mean (taking actual examples): three in a firm with thirty-five surveyors, or seven in a 'large' regional office of seventy, and only one or two in many other smaller firms, 'the smaller the firm, the lower the proportion of women' ... 'London is undoubtedly the best bet for women, if you don't want to be completely isolated'.

Table 2 Percentage of JO membership by areas, 1986 (male and female)

Area	Percentage
Glasgow and Edinburgh	7.0
Rest of Scotland	4.6
Merseyside and Manchester	8.6
Rest of northern England	11.9
The Midlands	9.5
Lincolnshire to Suffolk	3.8
London	22.7
Rest of the Home Counties	14.0
Rest of the south and south-west	11.6
Wales	4.0
All of Ireland	2.3
Total	100.0

Source: JO, RICS, 1986: 3. The situation in 1989 remains similar.

Range of contacts

I started by visiting one of the largest and most prestigious of the up-market West End practices in London, which comprises the full range of professional activity and a fair spread of different types, ages, and specialisms of women surveyors. They arranged for me to meet with about fifteen women from various sections. To balance this private practice emphasis, I then visited a group of women surveyors in one of the large insurance companies that employs a number of surveyors to advise on its property investments which are the backbone of much pension and insurance fund financing nowadays. I also contacted women who work in other non-surveying organisations such as large commercial companies with their own estate management departments, including industrial undertakings and nationwide chain stores.

In addition, I contacted women who work in other non-partnership structures such as the large building contractors. These have their own construction industry subculture which is somewhat different from that of surveying, and which some say is more favourable to women (few though they be in this realm), because it is more meritocratic and less 'male upper-class' or exclusive. Interestingly, some building and quantity surveyors within the RICS also see themselves as part of this 'other' culture. As one chap put it most forcefully, 'I don't belong to the landed professions, I belong to the construction industry' (a point replicated frequently by others, including women), showing the subtlety of the situation which might be lost on outsiders.

I was kindly invited by a very well-intentioned senior man surveyor to visit one of the main central government property agencies to meet both him and 'his ladies'. A considerable number of women work in the public sector, and therefore I also visited both the headquarters and the local regional offices of a certain major transport board with enormous property interests.

Not wishing to over-concentrate on London and the big practices, I visited a prestigious local practice, a large Midlands practice, and a truly rural one. I made contacts with women in the local estate agents in my own town to get the full range right through the different status levels and sizes of practice. I visited one of the main local authority departments in the provinces that employs an increasing number of women, especially at the trainee level, and made contact with a number of women in valuation and estate management departments elsewhere. I often met contacts I had made at these meetings again, in a range of other professional and social meetings which I attended in London and locally.

I was invited to meet with members of the Lionesses Committee (women surveyors association) early on in my research, and was subsequently asked to speak at one of their meetings. I was to keep coming across a core of women from this group at several subsequent events as the world of surveying is very close-knit. Some other women were of the opinion that the Lionesses were not representative, either because they were too London-based, or private-sector-oriented, or too 'up-market'. I tried to balance this by contacting, by telephone, a far wider range of women surveyors throughout Britain in all sorts of specialisms, stages of personal life, and levels of professional status, including those who were currently out of surveying practice.

I was able to contact a fair number of women ex-students bearing in mind that, until only a few years ago, five per class would have been an exceptionally large number. However, it seemed more valuable to seek a wider range of women to get more variety and coverage. Overall, there seemed to be little correlation between academic achievement and career success. It seemed that the personable, worldy-wise, yet 'average' woman was likely to do far better, and was less of a threat to the men than a precocious 'intense' woman.

I spread the net wide and managed to get a letter printed in the *Estates Times*, (15.5.87, No. 894: 10) asking women in property to write to me, and got several fascinating replies, including ones from that rarest of breeds, the woman property developer. All my respondents experienced the 'same' problems, whatever branch of the landed professions fraternity they encountered in their daily work, and whatever 'class' they were. I also kept in touch with members of the JO, and at their request sent them a short summary report of my research findings, which was tabled at one of their meetings. I was pleased when one of their leading London representatives sent me a letter stating that her fellow members agreed with many of my observations, including some of the young men, and thought it was 'so true'.

I contacted women outside the profession but within its influence. I talked to a few wives of surveyors, including those that had been in surveying themselves, who might be seen by the world as 'failures' in that they had gone into clerical work or turned to working with children. I found one particularly unhappy example of a wife of a surveyor who was renowned for his social awareness, but whose equally qualified wife had gone into ordinary clerical work out of boredom. She commented (with apparent approval), 'my husband's work on the problems of the inner city is far too time-consuming for him to think about women's issues'. (A touch of the *Stepford Wives*? Levin, 1974.)

Although I did not make a comparative study of men surveyors (like Gallese, 1987), in the course of my travels and conversations, I came across a fair range of 'incidental men', including ex-students. To give an idea of the disparity of progress, when I came across a male contemporary of one of the 'first women' who had been at college in the late 1960s and attempted to contact him, I was invited by his office to speak to him on his car telephone. He also appeared to have a personal secretary and office in the City and another personal secretary and office in the London Docklands. In contrast, my 'first woman' had an average but good position with a government body and was quite unobtainable when she was out of the office on site, but I could leave a message with the switchboard.

Other incidental men consisted of younger male surveyors, personnel officers, and professional section heads, all of whom gave me further ethnographic anecdotes. Government bodies are very conscious of their

Education and practice today

public image and the personnel officer might waylay me and give me his version 'before you meet the ladies', almost as if he thought I was some sort of inspector sent to check up on him. Another told me with sincerity, 'we believe in equal opportunities, 60 per cent of our clerical staff are women'.

Employment

Although increasingly large numbers of young women surveyors have been entering the private sector, many women are still found in the public sector. It is only in the last couple of years that the private sector has begun to overtake the public sector in recruiting women and providing the leading role models for women within the profession.

Table 3 Percentage of all young surveyors in each sector

Category	1986	1988
Private practice	53	56
Insurance and pension funds	7	11
Contractors	9	9
Others	12	6
Public sector	19	18
Total	100	100

Source: JO, RICS, 1986: 1; JO, RICS, 1988.

Table 4 Percentage of all women surveyors in each sector, 1988

Private partnership firms	40
Pension funds, insurance, etc.	10
Contractors	2
Public sector (including 2 per cent in education)	37
Not professionally engaged and other	11
Total	100

Source: Table 4 and related observations are based on figures from the RICS for 1983, 1987, and 1989, combined with my own estimates from the research. Although Tables 3 and 4 are not directly comparable, all categories, except public sector, are effectively private sector.

Few women surveyors are working part-time compared with women solicitors, of whom 26 per cent of those who were admitted in 1977 are now part-time (*Law Society's Gazette*, Vol. 85, No. 30: 5, 24.8.88; Marks, 1988). 'It's all or nothing, they don't want part-timers, it's a twenty-four hour job' (which it is not necessarily) (Molyneux, 1986). For further details of the situation of women in other professions see Appendix 2.

Progress of the young woman surveyor

I ploughed my way through a range of glossy recruitment brochures which the various public bodies and private firms produced to attract graduates in their direction. Much of it was classically 'male' in terms of descriptions of future careers (without breaks) and the photographs used. Some application forms asked the strangest questions as to weight and height and details of sports and interests. Some of the public sector bodies, however, have tried very hard to present good female role models in their literature. At the recruitment stage after leaving college (or more likely whilst there in the last year during the 'milk round' in which surveyors will come to the colleges, in what is nowadays a sellers' market), 'all seems fair' as the women and the men seem to get offered the same jobs. Nearly 100 per cent of all surveying graduates get jobs. Nowadays, around 60 per cent of all our students, but 80 per cent of female students, go to London, most to the large prestigious firms and this is fairly typical of southern colleges. They are likely to be motivated by the demands of training and gaining status, rather than by concern with particular urban policy issues, in their choice of job and location.

Many young surveyors move out of London after a year or two, but often remain in the south-east. Some women swap over to the public sector and appear to be much happier doing what they see as a wider and more interesting variety of work. But the private sector is in a period of expansion and many people have said if you go into the public sector that shows you are no good and a failure. This is very different from the past when around 40 per cent of all students would have gone into local authorities to get 'experience'. (Nowadays, less than 20 per cent of all surveyors are in the public sector, representing a continuing decline as in 1985 the figure was 24 per cent (*Chartered Surveyor*, 12.9.85, Vol. 12, No. 220: 668). It is now almost obligatory to go into the private sector in London to get a grounding, in the same way that it used to be considered essential to work in public service especially in the 'DVs' (District Valuation Office) to get a wide range of experience. People are unlikely to get the full range of experience in one London firm (often being 'stuck' in one specialist department) that they would get in local government, or in a small all-purpose private practice. This may not be a major drawback as the trend is for people to specialise more than was the case in the past. Many women have warned me of the dangers of being so good at one particular area that this actually works against them as they become 'indispensible', and they are passed over in favour of generalists (usually men) when it comes to promotion. A woman dare not be a generalist, as she must be seen to be 'good at something'. However, various men have commented to me that ability to attract clients (i.e. business) may count more than ability to do specific surveying work, and in the case of women, their 'sex' may to be their advantage in this respect, although their 'gender' may disadvantage them in other ways.

On taking the first job, there may be the initial shock of starting at the bottom again as the junior and being given fairly menial work even in a 'posh' firm. Several women have commented that they suddenly seemed much more alone, as if all the other women surveyors had evaporated into thin air. However, many ex-students who go to London share a flat with someone they knew from college and this softens the blow. What they actually 'do' at the beginning seems to vary considerably from (at worst) colour-washing plans to being given quite difficult structural surveys or valuations, and being sent out on their own with full responsibility for their actions. Much depends on personality, and it is difficult to generalise, as it is not necessarily the women that get the rubbish jobs. Indeed, the women may be seen as more mature, competent, and useful at the beginning, and the men are given time to catch up later. If there were a complete 50/50, male/female split in surveying, then I dare say women as a group would be allocated lower status jobs, but while we are at the stage of a few exceptional females among a sea of more average males, other rules apply (at least at the beginning of their careers). In housing, where there are relatively more women, it is already much more noticeable that a disproportionate number of women are in lower level positions (Levison and Atkins, 1987: 10, Figure G).

Salaries vary considerably, not just between male and female, but between different individuals and firms, although several women are convinced the men are earning more (one told me she was definitely paid £3,000 less than a male colleague). The JO did not find much disparity in salaries between the sexes among the younger age group (JO, RICS, 1986: 13), although those women who worked part-time (few though they be), proportionately earned less. Women who kept going above the age of thirty often received reasonably good salaries, but it should be remembered that there are relatively few of them. Contrary to popular opinion, in the midst of all the investment wealth of London, some young surveyors are considerably worse off than their colleagues in local government or the provinces. (Not all are earning £30,000 a year by any means, but significantly, a few new graduates do achieve £25,000 in their first job nowadays, *Estates Times*, 28.7.89, No. 1005: 1, 'Get-rich students cash in'). There seems to be pressure to dress well to fit in with the affluent image regardless, as one young woman put it, 'I only had the suit I had worn to my interview and I wore it for months on end, perhaps it made them take me more seriously as they thought I wasn't interested in clothes'.

Whilst at the first interview the 'treat them the same' factor may be at work, once in practice 'in the grown-up world' the situation may change if women seek another interview to change jobs. Several women have mentioned loaded interviews (occurring recently) in which they have been asked about their personal lives and whether they have children. 'If the firm is like that, I don't want them and they don't want me, I can always go somewhere else.' 'If a man gets married it's a sign he is settling down and it's seen as good for his partnership prospects, if a woman gets married she is taken out of the running and written off.' There are some firms which I have been told 'would never employ women' and so everyone 'knows' it is a waste of time applying to them, but these are in the minority, and 'you can always get a job elsewhere'. Indeed, one man bemoaned, 'we can't keep them, they can always go somewhere else'. There seemed to be a double standard in operation in what he said, as I have heard males being praised for their dynamism when they move. However, it is true that some of the more traditional firms still want people 'for life'.

Opportunities abound for younger women surveyors, but it is less so for older women as the pyramid narrows. When it comes to employers looking at the pool of potential candidates for further promotion 'it's almost as if the women don't exist' (although again, a few do succeed spectacularly). I would go so far as to say that many women are recruited on completely different criteria (although they may not realise it) to meet short-term 'man'power shortages, and that in reality employers have very different expectations of them. These factors might not manifest themselves for several years as all surveyors are 'juniors' until their late twenties (and they have to complete their test of professional competence before counting as fully qualified). I have come across many women who have sailed along, obliviously thinking that all is well and that they and their work are highly valued, when they get an almighty shock by being overtaken by men whom they had discounted as unimportant. The fact that the senior men seemed to treat them decently or even like them may have counted for nothing in the long run.

What do the men really think of women? One very senior man in a prestigious London practice told me, 'I think the chaps welcome more women being taken on, as they know that means there is less competition for them as the women are bound to leave. It's pleasant to have women around, the chaps like it'. Another comment was, 'We like to think we're modern. It's good to have a computer and a couple of women surveyors in the office, it shows we're with it'.

A woman lecturer in another surveying department was convinced that the old 'finders, minders and grinders' management concept applied adversely to women. Talking to a range of people, it would seem to me that women might be prematurely put into the first category when using their charm, breeding, and attraction to deal directly with clients (who are male of course), but as their looks decline they might become minders, or indeed more likely 'helpers' on other people's work, and eventually they might be channelled sideways to becoming mere grinders. They are being given these roles in a sequence that men seldom experience, their true ability bearing no relationship whatsoever to their income, status, or seniority.

Vertical distribution of women surveyors

Introduction

It is one of the main propositions of this study that in order for women to exert a strong influence on the nature of the profession and put forward alternative policies, they have first of all, got to be fully accepted into the profession and achieve positions of seniority within it – which in surveying would usually mean reaching full partnership level. Regarding vertical distribution, even allowing for the fact that many women surveyors are under thirty years of age and in fairly junior positions, there are nevertheless a considerable number of older women around who have failed to reach the levels of seniority to which they consider they should have been entitled. It would appear (based on figures given in strictest confidence by public sector bodies and extensive enquiries on the grapevine in the private sector) that very few of the surveyors at full partnership, or higher management level, are women. According to the RICS Records Office (1989) around eighty-two of the total number of women in surveying are at 'principal' level (compared with thirty-six in 1983) but I would estimate that barely a tenth of these count as 'really' senior partners or equivalent. (In contrast, the 2 per cent of women surveyors in education (Table 4) do proportionately better in terms of gaining seniority.) Included in this eighty-two are women who are partners in provincial practices. For example, one, who is a partner in her father's firm, said that she was always expected to take over the business because there was no son, but was initially resistant to her destiny: whether this counts as 'equality' is another matter. However, she now feels rather pleased that she became a surveyor, and is highly thought of by many. Interestingly, I have also met men who were told from their earliest years that they were going to be partners in the family firm, some of whom had purposely failed examinations at college in attempt to get out of it, 'it's not fair, women have the choice, men don't'.

Private sector

Many older women have reached associate partnership level, which in surveying provides a pool of possible 'talent' from which further promotion to full partnership is made. Women have commented to me that a log-jam is building up as the new wave of women surveyors reach their thirties and are not promoted further but left at this lower management level. One commented, 'Now I'm older I can see men who are younger and less experienced being picked out and promoted for partnership, whilst I'm still left where I was. They won't let me through'.

However, there seems to be a myth that there are not enough competent women around, or that they are not putting themselves forward, when in fact I have met many women who are blue in the face with trying. An alarming example of this view was expressed in Parliament (Hansard, 1986[1]), specifically mentioning women surveyors.

There was a general feeling among many women that men are likely to be groomed for partnership earlier, whereas women, after an initial climb upwards, are then shunted horizontally into specialist areas rather than continuing vertically, 'the men are all being given associate partnerships and bigger car allowances; we feel ignored'. Although a few women have broken through to partnership level, there are partners and partners. It can be a way of employing people on a fee-sharing basis (with taxation advantages) or it can be a sign of seniority. Even at the non-partnership level one comes across men surveyors who are employed on what appear to be very low salaries, but in addition they have the benefit of an 'attractive package'. Many women feel uneasy about this sort of arrangement, particularly if they don't understand how to play the game.

So far I have only found three of my women ex-students who have become partners (but know of a number of male ones). Ironically, in the later stages of doing this research, there have been a few spectacular appointments of women to full partnership level, but it is noticeable that some older women, who by rights should have been taken first, have been 'ignored' and much younger women have been chosen, creating mixed feelings all round. Having heard the life story of several of the women involved and their jubilation or bitterness, it is a very 'difficult' situation. So now women at partnership level seem to be composed of a few of the older 'exceptional' women and a number of much younger women who appear to have leap-frogged over others who were battling their way along: all this further 'fragments' women as a group. Older women who had left surveying (few though they be) and subsequently returned (perhaps in their late thirties, JO, RICS, 1986: 13) were also unlikely to regain a position of sufficient seniority to be considered for partnership; but again there were a few spectacular exceptions to this rule.

There are large numbers of small practices out in the provinces consisting of only a few partners, most of whom are male. Such practices survive on the assistance of the 'office ladies' they employ, many of whom are doing quasi-professional work. Even nowadays one finds that a few young men will seek to establish themselves in a 'new' sole practice. There seems to be another subuniverse within the patriarchal professions of the independent practitioner who is found in small towns where all such men belong to the Rotary, Masons, and the Lions, and are big fish in a small pond. They may be highly conservative and local in outlook rather than cosmopolitan or 'dynamic' (echoes of Pahl, 1965 and Gans, 1967). Many women have found, either as fellow professionals, office staff, or clients, that such men can be highly patriarchal in the way they treat women; but there are also certain individual men who are very helpful.

Some women have set up in practice on their own, but whether this should be seen as evidence of equality or exasperation is another matter. I came across just a few women building surveyors who were working on their own from home doing house surveys, combining this with their family commitments. Nowadays, with the problems of indemnity insurance and other overheads, it is significant that women are entering this area as men are leaving it. As one man put it, 'independent building surveyors are a dying species, it's too much worry nowadays'. Going into a technological area is not necessarily a sign of progress. Likewise, being a partner or owning your own business is not a sign of power or incipient capitalism at this level (it may be the result of sheer exasperation). Even women who set out with feminist intentions tell me their minds are perpetually preoccupied with 'keeping the business afloat'.

In contrast, solicitors and accountants are organised in a way which allows for many smaller firms as well as the larger London-based ones; generally they have a 'flatter' hierarchical structure with more professional staff being partners rather than employees, and with less emphasis on the intermediate band of associates found in surveying. Many women solicitors envy this, 'you're lucky, it's all or nothing for us'. Most legal practices now have at least one woman partner (Law Society, 1988: and see the *Law Society's Gazette*, 25.3.87, Vol. 84, No. 12: 921), although there is still a marked disparity in progress (*Law Society's Gazette*, 3.2.88, Vol. 85, No. 5: 3). The Law Society is now getting more and more worried about 'the recruitment crisis' (i.e. too many women) (Robinson, 1987). Fifty per cent of newly qualified solicitors are now female, and yet the profession is making very little allowance for their needs. Therefore many women are leaving in their late twenties and early thirties. One of the big problems for women, of partnership based professions, is that as associate or full partners they are not covered by normal employment legislation as they no longer count as employees and cannot therefore depend on getting maternity leave. Women solicitors (Law Society, 1988) have recommended that 'special' provision for women should be written into the Deeds of Partnership, and paid for by the fee-earning capacity of the whole practice. This has had a mixed reception.

Public sector

The categories of work are superficially 'the same' in the public sector but the ethos is somewhat different. Although there are a significant few who are in the higher echelons of public service, there are four women in particular who are at the highest level in one branch of public service surveying. There is then a considerable drop, without much in between, to the lower levels. One can see a similar jagged pattern to the 'splutter' effect in education (see p. 91). However, at the junior 'cadet' level, around 30 per cent and in some cases 50 per cent of new entrants will be women, and there 'appears' to be no discrimination; indeed, it would seem that they prefer women. I was told, 'we could assess the effect of equal opportunities better if women did not leave to have children just when they are at an equal stage of pay with men: opportunities are there if they want them'. In the public sector women may actually be better off than those in the private sector in terms of salary, pension rights, and equal rights. 'It is less exciting but more secure than the private sector.' The men are seen as dull and grey. One hears of 20-year-old males working out their superannuation payments for when they are 65: obviously they are not the risk-taking type. In contrast, many of the women are likely to be entrepreneurial in character and 'too lively and unsettling'. Being 'too bright' was even seen as a reason for women losing out on promotion in some organisations. Many women are convinced that they are given less challenging work to do, but it is scrutinised more by the men.

Nevertheless, the public sector is seen to offer women 'responsibility' and an interesting range of work, especially in central government departments in London. 'You can be doing anything and they have a policy of encouraging you to take complete responsibility. You are your own boss'. This is another paradox as in the private sector many women did not feel personally part of the enterprise culture, 'I was just doing all donkey work for my section head and he got to meet the clients and he got all the praise for my efforts. I am very glad I moved over into the public sector; they make you feel valued'. (So much for stories of women being lost in big state bureaucracies.) How long all this will last is another matter as everyone mentioned the imminent privatisation of government assets. One senior man told me, 'we are responsible for more millions of pounds than any private company, we are the biggest, look at all the Government property there is'. However, in terms of 'status', the surveying subculture nowadays puts the private sector above the public, regardless of the value or volume of work involved.

Women may be working in the centre of government but have very little opportunity to shape policy as against carry it out, as most decisions seemed to come from 'on high', but many women just accepted this, 'you're too busy to think about it'. Women talked enthusiastically of being responsible for new prison construction, Ministry of Defence property, and one surveyor was even in raptures at being in charge of the management of government property including unusual types of development such as 'dog pounds', which she found so much more interesting having previously been in a very boring job in a prestigious private practice. I have also had glowing reports of a rural surveyor lady who is responsible for managing the farm land around the perimeter of a certain nuclear power station. Some women are so proud of their work and love doing it so much that it does not even occur to them to question the debatable nature of some of the projects; but then that is not their role as civil servants. Likewise, others did not question the tax-collecting role of the district valuation office in which they were involved; but at the same time they espoused strongly conservative and entrepreneurial values.

Horizontal distribution of women surveyors

The extent to which the presence of women can affect 'what is built' is related to two factors: the nature of the specialism they are in and their motivation for going into surveying, in particular their awareness of urban issues. Regarding horizontal distribution, in general women seem to be increasingly channelled into certain specific areas, (parallels with the medical profession, Lorber, 1984). Three areas seem to be specially reserved for women although of course there is also a majority of men in these areas too (as there is in all areas of surveying).

First, anything that has an element of 'prettifying' in the worst and most superficial sense, whether it be in relation to conservation, landscaping, or creating an attractive image in the office by their very presence, seems to be reserved for women. Second, anything to do with residential development is a 'woman's area'. As stated, many women surveyors avoid 'resy' (residential) like the plague because of its low status association with estate agency (selling houses) which is a quasi-professional 'trade' area. This is unfortunate as many urban feminists are concerned about housing as a woman's issue, although one can understand why women professionals do not want to be marginalised into this area. More broadly, many architects would not even see house design as real architecture, presumably because there are so many women in it (Wekerle *et al.*, 1980: 205). This is also an area where lower status men might gravitate, especially those without the right contacts. I have observed that the rare male working-class student might actually say that he wants to go into estate agency as his first choice.

However, many women are initially attracted to surveying because they like the look of estate agency and then realise there are 'better' specialisms to consider. Other women were attracted to surveying because of an interest in 'male' areas of practice, and found themselves shunted into estate agency. In particular, some women with farming backgrounds were fascinated by auctioneering and land agency, 'I wanted to sell cattle and ended up selling houses'. Nevertheless, such women are still a significant group within rural practice. They possess the surveyors' love of 'getting out and about', but when this is manifested in a love of horse-riding, it is seen as a big joke by the men. Indeed, the word 'ponies' is met with the same reaction as the phrase 'women drivers' (a control on women taking over 'space' perhaps). In the past, it was also possible for 'ordinary office women' in estate agencies to work their way up to being a negotiator and eventually to be allowed to go on a quasi-professional course, and in some cases make the leap across to the RICS courses. In spite of the talk of more opportunities for women nowadays, this route has now been effectively sealed off, because of the emphasis on college education and the hardening of 'status' barriers in the profession.

The third area which women are seen as 'naturally' fitting into is 'property management' – not management in the executive sense, but in the sense of 'caring for' property almost as an extension of the traditional housewife and 'helpmeet' role. For example, such women will be concerned with keeping the tenants happy in a shopping complex development and dealing with rent reviews, servicing contracts, and public relations. It should be noted that in none of these specialisms are the women directly involved in actually planning, building, designing, or making development happen, nor in large-scale investment and risk-taking activities. All this is mainly reserved for the men. Another alarming trend in all specialisms which I have very mixed feelings about, although on the surface it seems 'good', is the increasing number of women going into property research. Paradoxically, women with considerable intelligence, expertise, and even feminist awareness give their brains to the service of male policy-making over which they may have very little influence and which is not necessarily in the best interests of women.

However, women welcome work that is related to commercial development, to avoid 'resy' (residential) perhaps? There is very little social content in all this as most of the work centres on investment, letting, and management of portfolios of office development. Indeed, it is one of the most capitalistic and patriarchal areas of surveying, but nevertheless many women want to enter this area (indeed some see it as 'glamorous') rather than question its impact on women and seek to change it. It is said that around 80 per cent of all the investment related to office and commercial development is concentrated in London and everything in the provinces is small fry in comparison. However, this development does affect 'what is built' over a much wider radius, generating housing, transportation, and service industry throughout the south-east and adding to the overall congestion.

Much smaller groups of women of particular interest are those who have reached positions of authority; those in the most male-dominated macho-technological areas of surveying; and those who are 'unusual' in some way, such as the very small number of black women surveyors. I came across about ten black women surveyors, and a very slightly smaller number of men in total, excluding those in housing management and overseas surveyors, in the course of my investigations. I found it difficult to find black women as those whose photographs occasionally appeared in the journals might in fact be 'only' secretaries (*Estates Times*, 21.2.86, No. 833: 10), but I did find a few black women surveyors by word of mouth.

These three categories of individuals often exhibit most strongly the essential attributes required of the ideal woman surveyor. However, even in the areas where there are very few women, mechanisms may still be at work to create a gender hierarchy. For example, one woman, a mining surveyor, told me that she wanted to be involved directly in mineral extraction, whereas her employer thought she should work on spoil heap reclamation, 'landscaping with trees and hedges, he thought that was more ladylike'.

Likewise, women that appear to be in technological areas may in fact be involved in legal contract work or cost management rather than the building process itself. Therefore, whatever specialism women appear to be in, one has to look very closely at what exactly they are doing in it, and for whom. Many women feel they are seen as 'helpmeets', there for the benefit of the men and not as people in their own right. Some women are involved in industrial property which superficially may appear more male, but the actual work they are doing may be similar to that related to commercial property, for example, rent reviews, rating and valuation work, landlord and tenant matters, and general supervision and management of property. There might be an element of glamour in dealing with the new science parks and high-tech sunrise industries, but I am told for valuation purposes the buildings are not that different from traditional industrial units: 'four walls and a roof, just a fancy shed'.

Some women seem naturally inclined towards technology. One woman confessed that she had liked bricks since she was nine years of age. Another woman explained to me that, when she was little, she used to go out with her mother who used to lift her over the site railings to look at the foundations of the new houses and tell her about construction (when the workmen were not there, of course). Whilst some of the women in the more technological areas work on their own or in small firms, several are with large contractors, for example, working on motorway construction, which involves 'dealing with the men'. Women who have got into the most male areas, and have established a specialism in this area, might find that their expertise in itself bars them from further progress or getting back into the office in terms of promotion. 'After all, most older men are no longer out on site, they have management jobs in the office, it's only a temporary phase for them'. Paradoxically, other younger women would claim, 'there is lots of encouragement to get on'.

Conclusion

I became aware, in my travels and conversations, of just how capricious and paradoxical the whole situation was. For example, many of the most successful women had entered surveying 'quite by chance'. Subsequently, however, some would find that there were surveying connections in their family, almost as if it ran in the blood. One woman had had a major row with her father and went off to London and entered one of the more male-dominated areas of surveying, 'to show him my independence'. Later her mother said, 'your grandfather would have been pleased' and she was told that he had been a surveyor. Women who thought they were acting independently, even rebelliously, might find that they were unwittingly fulfilling the requirements of the maintenance of the subculture. I came across many other women who were very much part of surveying dynasties or extended surveying families and who seemed to possess more 'cultural capital' than those who had entered from outside. In conclusion, it would seem that every woman's experience was somewhat different, but all had to learn in their various circumstances how to 'get by' in the world of surveying.

Education and practice today

Note

1 It was stated that the government simply could not find any suitably qualified women professionals, including planners and surveyors, to serve on the London Residuary Bodies. During the discussion Lord Elton was asked the difference between a man and a woman professional. He replied that he would hope the latter would be more attractive! (Hansard, 1986).

Chapter nine

Getting by in the world of surveying

Introduction

The overall ethos of the landed professions remains observably 'male'. When it was proposed that a women's centre in Oxford should be built by women, a cartoon appeared suggesting that monkeys should build a proposed new monkey house in London ('Stackup', *Estates Times*, 16.5.86, No. 845, back page). 'Oh, that's extreme, surveyors aren't like that!'

This chapter looks at 'all the little things' that women encounter in striving 'to get by in a man's world', (Riley and Bailey, 1983) These incidents are not trivial or personal, but the very building blocks of women's position in surveying, which may be summed up by the frequently recorded comment, 'we feel we're being scrutinised all the time'.

The office milieu

Many women have problems with the atmosphere of the professional office. The emphasis on sport and socialising which spills over into office conversations can make women feel left out, 'men waste so much time every lunchtime and after work, propping up bars and chatting. I can't waste my time like that, but they see me as disloyal if I don't'. Some offices have a fixed evening every week when all the men will go along to a local 'hostelry' and drink and chat. These are traditionally all-male gatherings, and it is difficult for the women surveyors to know whether to go, and how to behave if they do. The men likewise may be inhibited if women are present in social meetings and formal committee meetings. It would seem the sociable, personable woman fits in best within this environment.

The situation is more unpredictable in remote rural areas, 'they had never seen a woman surveyor before, they didn't know how to treat me'. One woman moved to one such area and when she realised that the men

Education and practice today

were 'simply embarrassed by me', she took the opportunity to take over and computerised their valuation department for them. (Whether this was a re-enactment of the helper role of 'there's nothing like a woman's touch to put a house in order' is another matter.) The fact that women were not meant to be good at computers or management was no doubt quite secondary for these men, who were still coming to terms with the possibility of a woman being a surveyor. In another rural area, one woman told me how a very ancient partner of the firm used to come up to her in the street and kiss her hand in the most *gallante* manner. Another woman in a minority specialism in the north said that, at her first professional meeting, the men were admonished, 'gentlemen there is a lady present' (that is, no swearing).

I received the worst accounts from women who were working in public service, particularly at the local government level where one would imagine the situation would be more enlightened. The general opinion was that one would be treated 'better' and that the men would actually be more open to feminist ideas in a traditional paternalistic 'conservative' private practice, than in a 'socialist equal opportunities type local authority, where there is the whole local government fraternity to contend with' (echoes of Levison and Atkins, 1987). Paradoxically, women told me how many government departments will offer lengthy maternity leave, accept job-sharing, and flexibility to transfer between full-time and part-time employment, but such glowing reports are usually followed by the statement 'nobody has ever done it in our section, of course, if you are away for a week you lose ground and it may affect your promotion'. It's all for that hypothetical 'abstract' woman again! I had several accounts of major battles going on about girlie calendars and offensive behaviour right in the midst of some of the most well-known 'equal opportunity' authorities in London, 'they never bother to check the borough surveyors department, of course, because they think it's all men'. It appeared that some local authority men simply disliked women. Comments such as 'why don't you go home and look after your kids?' were quite common. Women reported being 'commented on...this got to some of the women, one developed a rash and others just felt ill all the time'.

However, 'you can do much worse'. I found several women who had gone into surveying, particularly within the public sector, as a positive escape from greater problems elsewhere. I met two women valuers who had previously done science, and several other 'refugees' from the women engineer syndrome. Such women 'didn't want to be stuck in the laboratory all day' and disliked its factory floor atmosphere. One such woman had already been subjected to immense sexism at a white heat of technology type university. For example, one professor who had been instrumental in making major breakthroughs in 'birth control' would pick the few women out in lectures and say, 'now girls, this is the chemical used in the production of birth control pills, you will know all about that won't you?' She thought surveying was wonderful in comparison.

Work and home

One of the most frequently mentioned 'professional' problems mentioned both by those with children and those without was how to reconcile the demands of work and home. It would appear that around half of older women surveyors have children. Among the younger age groups there are many that seem to be 'holding on'. Everyone told me, 'there is a massacre in the late twenties and early thirties' as women leave the profession. This is partly in order to have children, but many other women were leaving because of a growing 'disillusionment' as they began to 'see' that all was far from fair. However, increasingly, others carry on working, and may 'leave it too late' either to have children or to change careers. It is still not seen as natural to have children in the world of surveying, and if women do, it is likely to be seen as a purely personal and avoidable matter that is of no concern to the profession, 'you don't have to have children, you bring it on yourself'. One woman student declared, 'I suppose it's the fault of biology and not surveying'. Her whole manner suggested an inner conflict that was tearing her apart, for which there was no 'space' for discussion within her college. But I have noted that some women who have slightly older children are seen as less threatening and trustworthy because they have, 'accepted their role'. Again it is a matter of caprice as to which attitude prevails, 'if you've got over the messy phase of having babies, they know you're not going to leave and let them down'.

There still seems to be a lack of sympathy in the landed professions as a whole towards married women having jobs, let alone children as well ('Couples are out of order', Stone, 1983, a classic): although some individual men surveyors are most aware of the issues. One woman told me that I 'must' include the following example. In a certain provincial practice, a woman surveyor announced she was going to get married. It happened at that time that the headed notepaper was being reprinted. She was astonished to find her name left off the list of associate partners at the top of the page, 'oh, we thought you were going to leave, women always do when they get married'.

Although 'the men do both', the whole atmosphere is far from 'girl-friendly' (Whyte *et al.*, 1985) for women who try to have a career and a family; indeed, there were strong elements of resentment even misogyny beneath the surface. I was told of one woman who had got on very well with her male colleagues and then she got pregnant. 'It was as if the men felt she had betrayed them. They were frightened as if it had happened to them. It was as if they thought it would never happen to her, they said, "we thought you were different"'. However, some more 'mature' men surveyors are much more aware of the problems women encounter, especially if they have working wives.

The wise woman surveyor who is 'her own woman' will know 'it's all a male game', and will have children and a career. It would appear that those women who succeed the most in their surveying careers are also those that have been the most successful in producing and 'organising children'. Such women are often seen as 'superwomen' by other women surveyors, and seem to be blessed with boundless energy and willingness to do whatever is required, 'moving heaven and earth if necessary'. They never seem to get headaches, or aged relatives falling down stairs, or children getting sick, or if they do, no one ever hears about it.

One woman building surveyor was doing a fifty-mile round trip to site a few weeks after having her first baby; another woman had tried unsuccessfully to take her baby on site with her. Another woman surveyor told me how she and her husband moved next door to a school when their child reached school age 'to make it easier'. Also, astonishingly, I came across several women who commuted over 100–200 miles per day by train, leaving their children in the care of nannies or relatives. However, contrary to popular opinion, many cannot afford nannies, or if they are 'rich' it is only because they work and now dare not stop.

In contrast, the Institute of Housing has initiated a series of projects on women taking career breaks, returning to work, and attempting job-share (Leevers, 1986; Levison and Atkins, 1987). However, this must be set against a background in which women are making little vertical progress in spite of increased numbers and better conditions of service. Likewise in town planning, women are concentrated chiefly in the lower grades (RTPI, 1984, 1987, 1988).

It is sobering, among all the 'glamour' of the image of the successful woman in the professions, that several woman told me that most of the successful women in surveying had got there through personal misfortune! Divorce, in particular, is seen as 'the biggest occupational hazard in surveying' (greater than machinery on building sites presumably).

Dealing with people

Introduction

Over the centuries a whole series of class and gender relationships, collusions, and pay-offs have been developed between different types of people, upon which patriarchy and capitalism are dependent. A surveying practice, like any other office or organisation, will manifest these historical relationships. Women surveyors are thrown into the pot as a new ingredient that has to be 'dealt with' (echoes of Hearn and Parkin, 1987, and Crompton and Mann, 1986).

Typists

Women in offices are traditionally seen as 'helpmeets' and as 'attractive'. Regarding the model it would seem that there is a strong element of 'collusion' and 'colonisation' between high status men and low status women. The whole system is based on the idea that the women are there to give support to the men. Therefore it is hardly surprising that many women surveyors have said that their greatest problems at the beginning come from typists and other women office workers. Not only do they get no 'support', they may get outright opposition. This is operationalised through women's own personal feelings rather than through commands from the men. One woman said older women office workers felt threatened, even jealous of her, as if she was trying to take their bosses away from them, and therefore they were very uncooperative towards her. I had several examples of women finding that their typing was not being done, or done more slowly, all of which was an additional burden to impede their progress. Some women find it is the older women who are the worst, whilst others have said it is the younger women secretaries that are the problem. Even strong men have been reduced to tears by 'stroppy typists' ganging up against them so it is not always just a gender issue. 'If they are rude they are likely to be rude to everyone, but that bit more to the women.' This is further developed in the typing pool situation in local authorities where staff are shared, and problems may arise when women surveyors find their work 'will have to take its turn, as Mr Smith's work is very urgent'. If one is tough you go down (it's always down) to the typists and have it out with them once and for all. If a woman surveyor is nice to women typists she may be seen as 'queer', in particular if she has her own secretary and is friends with her. I have had several accounts of this situation. Needless to say, some men surveyors do treat their women secretaries badly, as evidenced by the publication of a letter from a secretary entitled, 'Pay for a dogsbody human being'(*Estates Times*, 29.3.85, No. 789: 10).

One woman said she found her desk relocated out in the main clerical section 'to keep an eye on the typists to make sure they are working'. This put her, and them, in an impossible situation. I was frequently told that I must mention the fact that a certain woman surveyor, who is generally accepted to have reached the highest level in a certain prestigious practice, has a male secretary. Some women get on very well with their secretaries and typists. 'I'm going on holiday with my boss', boasted one secretary to her friends. Her boss was a lady surveyor and they had both been to university together but one had chosen to become a secretary and the other a surveyor. 'IT COULD BE ME SITTING BEHIND THAT TYPEWRITER' stressed the latter.

Many women consider it unwise to 'fraternise' with the typists even if they are the only women to talk to, and one should never dress like them. Women have to constantly establish their 'different' status and avoid being 'read' in the wrong way, 'Hello, you must be the new secretary'. The general view was that once they got used to you and remembered who you were this sorted itself out, but one would still get the occasional 'stranger' who did not realise.

Keeping up appearances may entitle a junior surveyor to her own secretary in the more classy West End practices. 'Do you know my secretary was quite resentful of me until we chatted and we worked out she was earning more than me? I think the firm expected me to take any salary just for the privilege of being there'.

Technicians

Technicians are another potential problem. Some may try to be clever and test a woman on some obscure technical trivia that most men surveyors would not know either, and which only technicians are meant to know in order to do their job, 'what, you went to college all that time at the taxpayer's expense and you don't know how to use a planimeter' he said to me, waving a pre-war brass model in the air that was probably last used when working out field areas for the Enclosure Acts. Technicians think that if a woman cannot type or know all the skills of a woman draughtsman then she knows nothing about planning or surveying; no professional man would be judged in this way. The opposite may happen regarding the way 'ordinary' women view you and they may be disinclined to talk to women surveyors about domestic matters, 'oh, I didn't think someone like you would know about that sort of thing'. Another point many women have mentioned is a problem with certain security guards and commissionaires who, even if they know them, will always stop them as if to impress on them that they do not really have any right to be there.

Clients

There may be confusion in the eyes of the client as to who is the surveyor. Several women have told me how they have taken a male surveying technician along with them to hold the other end of the tape measure – everybody talks to the man and not to the surveyor herself.

Some do talk to the woman...whilst looking at the man. Some technicians who are helpful will play along with this in full agreement with the woman, enabling her to spring the surprise on the client at the end that she, not he, is the surveyor. Indeed, the surprise factor works well in disorienting the client while there are still relatively few women, and can be useful in negotiating terms, or 'helping to control a difficult client'.

Some men surveyors will take a woman surveyor along to defuse a situation, 'I knew he wouldn't be rude in front of you'. The men are using the woman as a shield to protect them from an ugly situation by appealing to the need of the client to keep face and not behave in an aggressive manner 'in front of a lady' (echoes of Pahl, 1977b: 147). Some clients, most of whom are men and many of whom are of an older generation, have some difficulty accepting a woman as the surveyor. 'I don't want to see the secretary, I want to see the surveyor.' When on the phone, a woman may have to repeat to the client many times that she is the surveyor and not the receptionist. Some will take advantage of the situation and innocently get information that they would never get in their role as a surveyor (provided they do not go against their professional ethics). One way round the question of establishing credibility as a surveyor with existing clients is for the man currently dealing with that client to take the woman along with him the first time and say, 'this is Ms Theodolite who will be taking over from me on this project and I have briefed her on the matter'. This also happens in firms where the woman is the daughter taking over clients from her father. One wonders if this would be necessary with a man taking over.

Sometimes other men make assumed connections between men and women surveyors when there is none, seeing them as wives, daughters, even mistresses or secretaries. One woman said she entered a clients' meeting at which she was going to be the speaker to present details of a proposed development, and was immediately commandeered and bossed around to set up the room for the speaker! It can even happen to a man 'with' a woman. A very young looking 20-year-old male student and a woman mature student went to visit a certain housing body and before they could explain who they were, they realised they were being 'assessed' for possible accommodation and were assumed to be mother and son. All this sort of thing is an additional burden, especially for young women setting out on their careers.

However, once they all get over the first meeting, and the client builds up trust and confidence in the woman surveyor, word gets around that 'she's OK'. But it only takes one woman to do it wrong and she may queer the pitch for all other women in that area. In all this the 'caprice' factor is in evidence, and you never quite know what social construction will be put on you in a particular situation.

Education and practice today

Ethnic minority women

The problem is even greater for ethnic minority women, especially black ones. Several such women have themselves told me, with amusement, that white men 'like' them, especially if they are attractive and 'not too dark' (*sic*), because 'we add a bit of colour to the office'. However some white men are not so keen on sending them out to meet clients, who 'expect someone white'. Some black women believe they are more 'acceptable' than black men because they are less of a threat, and therefore black men in surveying are more likely to be surveying technicians than professionals. One personnel officer in a local authority estates department told me he had no black surveyors ('of course') but referring to a district office in an ethnic area he said, 'but we do have one black technician [male] as we had to have one there'. I did come across two ethnic minority women, one in the private sector and one in public service, who have reached very high levels; indeed, the latter was one of only four women at her level, with over seventy men under her. The one in the private sector said that there are about 250 main people in the property world in London that one has to get to know, and it would seem once one was 'in' on this system, one was half-way to being accepted. She was obviously from a high status entrepreneurial background, got on well with everyone, and found little discrimination. Others had more problems. An article on 'Racism within the legal profession', written by a black woman solicitor, was most enlightening for me regarding the mechanics of the current situation in the professions (Amoo-Gottfried, 1988).

The building site

The big bugbear is the building site and the attitudes of the workers, 'the uncouth, uneducated, ignorant ones'. (Their role is complementary to that of the male professionals in maintaining patriarchy, they are not operating in a vacuum.) They may try various tests on women, and be awkward/rude/indecent to make the women feel unwelcome. Women have been tested in a variety of ways, by a range of people, for example, by being sent up tall buildings when they are interviewed for the job to see if they can stand heights, being expected to enter and inspect a building that is clearly unsafe 'with seagulls swooping on me', being expected to have impossibly detailed technical knowledge, and, for comparison in a rural situation, being expected to catch calves in the cattle market; and generally finding 'men really putting you through it'. However, once the men get used to the fact that they are not going to scare a woman away and they realise she knows more than them, then they may with sullen resignation, or even secret admiration, accept their 'lady gaffer' (parallels with the 'bossy woman' in the student group example), and even defend her against insults from other gangs of workers.

Sometimes workers and technicians take out their resentment towards the firm itself on what they see as the weaker members of the professional staff, that is, the women. Some may even try to show off to other staff members 'beneath' them, such as cleaners or porters, by trying to laud it over women surveyors, preferably when nobody else is present. One woman was spoken to most rudely by a technician when nobody else was around, in front of a new junior employee, as if to demonstrate that his male ego could not stand a woman being above him. If such matters are reported, it is likely that the woman will be told she asked for it and must have done something to upset good old willing (deferential) 'George'.

Negotiating

An issue of great interest is how women deal with other professionals, both men and women. Property management and investment can involve a fair amount of negotiating as to rents, prices, agreements and coming to a workable arrangement on a whole range of matters. When men meet (based on what men have told me) they are likely to spend a while discussing the weather, cricket, women, cars, etc., and then almost as an afterthought say, 'my goodness look at the time, let's see what I can do for you'. There then follows a prolonged period of competitive discussion in which both sides want to keep face and protect their ego. Men tell me that men always like to haggle and there are unwritten rules about offering high unrealistic figures first to protect the pride of each side – 'Men have always got to have the last word and win, or choose to concede'. If a woman and a man are negotiating, there is still the element of surprise and the likelihood of the man underestimating the woman. Women may also be much more direct in putting their final offer forward first, which may completely throw the man to the woman's advantage. Many women do not choose to go through the preliminaries about the weather, etc., and indeed some take a while to realise this is expected. Others read up about the cricket scores beforehand and play the whole game according to the male way of doing things.

I always asked women surveyors what happens when they meet another woman surveyor. 'I haven't had that happen yet' is a common reply as there are still relatively few around. Those that have negotiated with a woman say that it is much more straightforward, far less posturing and there are no male egos to protect. 'With a man you must make him think he has won to protect his pride even when he hasn't, let him think he has outsmarted the woman, or alternatively that he has been *gallante* to her...it all depends on his disposition...you have got to be a good judge of character'.

When women get together to negotiate, they can settle a matter much more quickly, so much so that they may stay in their meeting longer than necessary to give the impression they are doing their job thoroughly. Also they tend to have their social chat after and not before the deal is settled. It is not always sweetness and light, as some women find other women hard and competitive in business. When a woman tells me that another is, 'a real hard bitch to deal with', when the woman in question has previously told me how supportive she is of other women in surveying, one is in somewhat of a dilemma as to who to believe (but one divulges nothing to either party). If there are very few other women surveyors around, and the few there are are on opposite sides, this can further fragment women socially. The solution to this is to be like the men and completely divide personal and business roles, but several women have said this is much more difficult for them to do, 'it doesn't come naturally, and women are so used to giving way and making do that it is very difficult to stand one's ground with a woman in business...and then go and play squash with her afterwards, women still take everything much more personally than men'.

Marginal women

Another issue requiring a short digression is that of the role of Jewish businesswomen, (which I have been told I must put in by several such women, although others consider attention should not be drawn to this issue). Those seen as 'marginal women' (although they are few, they are high achievers and therefore appear prominent) often act as trail-blazers and innovators in opening up new areas for women which at first are seen as too 'trade' or 'hard' for mainstream women, but which they later take over as 'quite acceptable' and they are marginalised further. They seem to be judged by a double standard in which their business methods are scrutinised twice as much as in the case of mainstream men (or women). Women (or for that matter men) who are successful can suddenly find the ground cut beneath them with statements (from men, and even some women) such as 'well, they're all like that, the women are the worst, the family came over before the war penniless, the old man was in the rag trade in the East End you know'. (Parallels with the 'she was only a typist' example earlier.) The fact the woman in question may come from a long-established, non-business family, and has fought hard to have a career of her own is irrelevant (compare *Chartered Surveyor*, 4.12.86, Vol. 17, No. 10: 1036). One gets wind of similar things being said about Asian men who are going into estate agency, but few, as yet, have gone into surveying itself.

Presentation of self

Clothing

What women surveyors look like and how they behave matter, as they are judged accordingly. Different age groups of women surveyors have been subjected to different and even contradictory expectations as to dress. The contrast between some older women surveyors who 'learnt' to dress asexually and be equipped for all weathers in sensible clothes and the younger bourgeois feminists, with their business suits and high heels, is startling, but they are all surveyors, and are all trying to 'get by'. 'I can never be the real me.' The whole area of dress is a minefield with every sartorial manifestation from mini-skirts to waxed Barbour jackets, tweedy suits, grey business suits, pearls and pussy cat bows, through to jeans and men's waterproofs and overalls; all these being accepted, at some stage, as to what a woman should wear, all giving out different messages regarding the particular subuniverse of surveying to which they belong and also age, status, and personality of the individual concerned.

> 'Whatever I wear someone will make some comment or other', 'I never know what to wear it is impossible', 'I wore this very smart blue dress, "where are you going, anyone we know?" they said. I stick to a business suit now it's like a uniform and saves all the hassle of deciding what to wear. But when I get home I always take it off straight away as it's nothing to do with my real personality'.

All this can have damaging long-term effects, as pointed out in paragraph 3.9 of the *Report and Recommendations of the Working Party on Women and Planning* presented to the Institute's Council on 1 April 1987 (RTPI, 1987): 'Generally women felt they had to underplay their femininity, i.e. to some extent suppress their personality, in order to avoid casual remarks which undermine the ability to act as a professional equal'.

Many women found it quite 'difficult' to make the transition in dress and manner from college where they were meant to be the 'same' if not 'asexual' (and wear jeans), to the office situation where they were suddenly meant to 'come out' and be attractive, if not sexy, for the benefit of the firm – 'attractive women make unattractive property more attractive'. This attractive image brings its own hazards. Whilst there is undoubtedly still the occasional surveyor who will say 'I can see the light between your legs in that skirt' (as one young woman surveyor was told – and much worse), many men would be less likely to make such comments nowadays. This is not to say that I have not come across women who have experienced actual harassment within surveying offices, as the problem undoubtedly exists.

The current emphasis on the business suit does not go well with the essential anorak and wellies that many surveyors wear to go out on site over their indoor clothing. 'It's almost like trying to be two things at once.' Do women really have any choice as to what they wear? It is rumoured that in some Mayfair firms women get clothing allowances and a certain image is cultivated that goes down well with international clients. 'They only take some women surveyors on as they are very attractive and have the right accent. You can send one of them to a business breakfast with an American client at the best London hotel and they can carry it off beautifully. Some even speak Italian or French which is ideal for foreign clients.' However, in fairness, another man concerned with graduate training in a large and ancient London practice, said they would avoid this and use a man, as it was demeaning to think that women had to have some other redeeming feature, such as looks or linguistic ability as well as being a surveyor, to be accepted. In general, it still seems to be assumed that women will leave surveying before their beauty fades or when they have children. Otherwise they may find themselves shunted off into a siding and not groomed for partnership which is very unfair as they no doubt have had to work very hard to get thus far. 'Nobody wants a woman over thirty.'

In the provinces there are still many 'jumpers and skirts offices', and it has never occurred to the women, or indeed to some of the men, to wear a suit. One woman was quite surprised at the very idea and told me she had been issued with some men's waterproof overalls for site work (which were too big) and said, 'I suppose that's my business suit'. In planning offices and housing departments it is more difficult because of the slightly more socialist emphasis and the dislike of 'stupid middle-class women' who may be seen as 'Thatcher clones' (i.e. bourgeois feminists in a derogatory sense) if they dress smartly; but will be seen as hippies by the public if they dress 'ecologically' and as typists by everyone else if they rush out to Marks and Sparks and grab the first jumper and skirt they see.

Self-defence

Women often mention the issue of 'safety' after discussing 'clothes', no doubt because if one wears 'the wrong things' one can be seen as 'asking for it'. An issue that has arisen many times in conversations is the case of Suzy Lamplugh, the estate agent who went missing at the time of going out to show a client around a property in London. Although it has not been established what exactly happened to her, everyone talks about the case in relation to the problems of being attacked. What saddens me is that her reputation is obviously 'highly respectable', but newspapers such as the *People* (1.3.87, No. 5478(P): 19, 1.3.87, 'she loved woodland games') have sensationalised the whole business, creating a totally false 'sexy image' of her. This sort of reporting suggests to the 'people' that women surveyors and estate agents are of questionable morality and 'exciting' in the worst sense.

I was surprised by the intensity of feelings of anger and fear of violence just beneath the surface in some of the comments women made. There seems to be a range of opinions as to what women should do about all this, for in the 'out and about' profession, one is likely to have to go anywhere: empty buildings, private houses, isolated rural estates, etc. One woman told how she went to value a property, an isolated house which contained 'two mad brothers with an axe in the garden'. Women try to avoid being locked in, 'not that it would ever happen but it might', and always check on the exits before doing valuations or structural surveys. Another said she goes to the door of the premises and if she does not feel at ease she will say 'sorry, wrong address'.

On the question of self-defence some take the view it will only make it worse to defend yourself or to struggle, 'after all if you are going to be attacked it will happen, and it could happen anywhere, not necessarily on site' – a fatalistic attitude prevails that does not tie in with the otherwise very modern attitude of taking control of one's destiny and being a successful professional woman. In several areas, women and their firms are most careful to make sure that they report back after going on site. In some cases the practice of going in twos has been advocated which is not seen as an economical use of 'man'power (and some women fear this will be used against them, although men will only go some places in twos!) In one provincial town, the surveying and estate agency firms have co-operated, together with the police, in the running of self-defence courses for the women, but the reaction is mixed. If such problems are given too much prominence, will it put men off employing women; indeed, is it just confirming pre-existing assumptions? Also, if women have disassociated themselves from the unqualified women who are in estate agency, do they want to mix with them in a self-defence course where everyone is a 'just a little defenceless woman' all in the same boat? Some prefer the personal self-sufficiency approach, and may even see the measuring rods, rules, and other surveying implements they carry as having another possible role for self-defence, 'not that it would ever really happen of course'.

Some men's perceptions of women as 'asking for it', or assuming that a woman on her own is 'up to no good', are quite alarming. Male surveying students are usually quite sensitive in this respect, being fairly sociable types that 'like women' and may have girlfriends on the course; but just occasionally a male student makes an inappropriate sexist remark and one wonders how near beneath the surface primitive assumptions lurk. A group was presenting a project related to the redevelopment of a built-up inner urban location which had a reputation for crime, mugging, and vandalism: (my comments are in brackets), 'if women will walk across open fields they are asking for it (what open fields?), it's like (it's not?) with blacks and gays if they parade themselves in broad daylight they can expect to be picked on'. This remark produced boos and hisses from most of the group, but some agreement also. Women do tend to get lumped together with other groups as the cause, rather than as the victims, of crime. The exact opposite of Suzy Lamplugh, which also reflects this assumption that women should not be on the streets in some areas, is the account I heard from several women, of two women surveyors who were picked up by the police for prostitution because they were seen to be entering empty buildings with men. The men in question were property professionals meeting these women surveyors for a series of site visits in relation to a redevelopment scheme in a run-down area. It is probably better to look sexless and wear an orange anorak to avoid such misconceptions, but the client and the employer may see that as scruffy and expect a woman to dress smartly for the London office. Clearly women cannot roam freely through space in the same way as men (Cockburn, 1985b), even if their professional work is the management of space.

Cars

The development of the motor car constitutes the ultimate mechanisation of the domination of space; therefore the size and make of the surveyor's car have all sorts of symbolic meanings to men – it is an outward and visible sign of their inward professional status and seniority. Many women choose not to go along with this and may actually be reprimanded by their firm. One woman said she likes something small and practical as it's so much easier to park when going out on site. Another said she liked her second-hand car and did not want to use the firm's car allowance to buy a new one. Many women have commented that they and their colleagues liked small cars, although they did concede that they might be persuaded to go up to a quality medium-size car, but were not very keen. There is much pressure in the private sector to give a good image by having a smart new car.

Although women themselves are not too concerned about cars, they do notice that some men of the same age as themselves suddenly get larger cars and bigger expense allowances, almost as an outward sign to tell the world that they have just been picked out for priming for partnership (about 70 per cent of all young surveyors have a company car or car allowance, JO, RICS, 1988: 36). Perhaps in a man's mind the fact that a woman does not want a larger car is clear evidence to them that she does not want partnership and she is not really serious? Men play by such strange rules. In contrast, another woman told me that when she worked with a government agency, they received a memo telling them that if they had two cars in the family to use the older one and 'to look poor' when they visited sites or went to value (i.e. tax) buildings, as it was not diplomatic for public sector officials to look too well-off (this could be destructive to a woman's self-image and 'authority').

An attitude of 'poverty' is also expected of lecturers by some students, 'lecturers are always moaning about their salaries but they can afford home computers' (in the private sector and in many other colleges they come with the job). It is considered good for students to boast of what they will earn in practice but are we selfish if we want to be treated like other professionals, particularly if we are women? One could argue to a point that it is our efforts that give them their fee-earning capacity in the first place? Paradoxically, a career in housing is sold unashamedly as a way of doing good and making money at the same time, 'taking one's fair share' when in fact many working-class people (the 'subjects') deeply resent the affluence of those who make careers out of their misfortunes.

Word-processors and computers

Education seems to be ahead of practice in providing computers for students' use (if not for staff) as I have come across several offices that do not have computers, including some valuers in the public sector and some quantity surveyors in the provinces who are still doing all their work by hand. Word-processing is a double-edged sword for women, and it is best not to admit that you can type, either in a student project group or out in practice where reports need producing, as you may end up being imposed upon, 'just this once, it's a rush job'. In contrast, several women I have met, who did not even study computers as part of their course at college, have built up expertise after they left college, developing computer systems for their offices and writing their own programmes. This would be quite an achievement, 'even for a man'. But some men do not appreciate such women, 'when I'm sitting at a keyboard working on a programme, if anyone walks past me they naturally assume I'm a typist'.

Education and practice today

Conclusion

Many women feel they are carrying an extra burden compared with the men, because of all the factors illustrated above, 'you're not given the benefit of the doubt, you have to prove yourself all the time'. In conclusion, from a woman surveyor: 'We have learnt to live in a male world and to adopt male professional work patterns, and to fit in and to learn their rules. The men haven't had to change at all, it's all been one-sided'. From a man: 'It has worked all right all this time, why have people like you got to come along and try and change it? We don't mind letting you in, but not if you want to alter everything. You don't understand it all, that's your problem'.

Part four

Implications for the built environment and the profession

Chapter ten

The influence of the subculture on what is built

The commercial emphasis

The market-oriented emphasis, present nowadays throughout the surveying subculture as evidenced in the pages of the *Chartered Surveyor Weekly*, and more broadly within the development fraternity as expressed in the *Estates Times* and *Estates Gazette*, is quite overwhelming. It would seem to preclude any 'space' for consideration of women's needs, or any other social needs for that matter. The presentation of land use and development schemes is 'so matter of fact', as if everyone welcomes and agrees with the proposed developments (presumably because of the profitable returns they bring), that one has to overcome a certain inertia not to fall into feeling that one is just 'being silly' in imagining that there might be another viewpoint.

But surveyors do **know** (and some care) that there are alternative viewpoints towards property development. For example, there is a cartoon (*Chartered Surveyor*, 30.1.86, Vol. 14, No. 4: 252) which has a surveyor saying, 'Of course we know about the community, we want to make as much money out of it as possible'. Indeed, one property company has as its logo 'who cares, wins'. Several large firms of surveyors now have their own research sections looking at 'social trends', albeit from the viewpoint of market potential rather than social policy. People who spot 'social need' and invest in it are admired, for example, owner-occupied sheltered housing for the elderly is a growth industry (*Chartered Surveyor*, 25.7.85, Vol. 12, No. 4: 250). Of course, women are not generally seen as a viable market in their own right in this pursuit of meeting lucrative social need. However, in the recent past some development surveyors involved in planning do seem to have been more sensitive, and even guilt-ridden, about their public duty to be aware of the needs of others less fortunate than themselves, 'nor are most planners educationally equipped for the new tasks...they are not trained to listen, to put themselves in other people's places' (*Chartered Surveyor*, October 1972, Vol. 105, No. 8: 160).

Effects and implications

Even when one looks at the more socially aware types of men surveyors, such as those involved in housing management and town planning, one feels that many are not completely in tune with the needs of the people they are meant to be exercising authority on behalf of, because they simply do not 'see' women. There also are still strong echoes of policy-making being done 'for' the working class, as 'we know best', that is, a top-down colonial approach. Surveyors, as professionals, traditionally must not get too involved with the people for 'the surveyor must be a servant of the community and not any part of the community' (*Chartered Surveyor*, December 1975, Vol. 108, No. 6: 121). Women as both the 'planners' and the 'planned' often feel a distinct uneasiness with the 'separation' between the professionals and the people (i.e. the public and the private realms) and many 'frankly find it unproductive and divisive', and therefore get their role 'wrong'.

In fairness, some surveyors do contribute their professional expertise on a voluntary basis to help tenants, as evidenced by the 'Pull Out Special' produced for *Community Action* (a radical housing magazine, Bread 'n Roses, London) by SIFT (Surveying Information for Tenants, for example, No. 80, Summer 1988, on heating and condensation). Several women (not from SIFT) told me how they had offered free technical and professional advice to people on a certain estate, without in any way seeking to 'control' the recipients. However, male professionals sought to move in and take over both the women surveyors and the tenants, and re-establish a patriarchal model of professional service delivery (one woman impressed on me strongly that I must put this in). Women may find 'their' design problems 're-presented' by male professionals as highly complex technical issues and taken out of their hands (Ware, 1987).

When surveyors seek to solve social problems, they tend to see the world in terms of land uses and buildings. When I ask some surveyors in the course of conversation, 'but what is reality?' the instant answer will often be, 'bricks and mortar of course'. In a letter by a socially aware male surveying student, 'Town planners may be right', (*Chartered Surveyor*, 3/10.1.85, Vol. 10, No. 1: 3) he pointed out that this spatial world view is a characteristic of right-wing thinking, even when it is manifested by the apparently more 'socialist' surveyors which leaves the status quo unchanged. For example, an article identified fifteen design faults as 'the problem' that caused the 'riots' at Broadwater Farm estate (*Chartered Surveyor*, 24.10.85, Vol. 13, No. 3: 297; compare Coleman, 1985).

Where are women in all this? 'Women? – that's not a land use issue'. Women and their needs are not seen as a special issue worthy of attention, 'women, that's natural, so it's not a problem'. But in contrast, I have frequently observed that surveyors will sincerely put all their energies into helping non-gendered, non-political 'worthy causes' related to 'feeding the world', 'Red Nose Day', or getting more kidney machines, etc., usually on the basis of being sponsored to indulge in some of their favourite sporting activities.

Land or design? A paradox

Following the 'illegal' demolition of the famous Firestone art deco building, Leslie Ginsberg, the architect, said that 'the greatest threat to conservation is the RICS whose members are trained to see their work as development portfolios', but it is to the surveyors' credit that this statement was published (*Chartered Surveyor*, October 1980, Vol. 113, No. 3: 148). A major paradox, or knot in the model, is that in spite of all this emphasis on 'land' and 'space', both students in college and many surveyors in practice seem to have a distinct dislike of 'design' or architecture for its own sake, which is seen as rather 'arty-farty', or 'too detailed'; or 'social', and therefore not 'real'. It is something which is 'imposed on the market'. So again 'land' is venerated as an abstract subcultural concept at the meso level and seen as an almost divine 'absolute' at the macro level, but the nitty-gritty of land use and development design as a spatial reality at the micro level may be seen as a 'nuisance'. Paradoxically, surveyors put great emphasis on 'the site', which might really mean 'the financial return from the site' rather than its physical design.

Even more paradoxical, in the more socially-oriented branches of surveying, especially housing and town planning, a concern with 'land' and spatial issues is also given low priority, which I find quite incredible as so many policy issues have a direct physical element. It would seem that certain of the more 'left-wing' practitioners, and academics in particular, are still going through the anti-spatial phase which afflicted geographers and some planners several years ago (as a result of the influence of neo-Marxian urban theory). Although 'resy' (housing) is the largest land use, normally comprising 70 per cent of all development in any town or city, it is seen both as 'down-market' by the more 'right-wing' private sector (although profitable), and as trivial and bourgeois by the more 'left-wing' sectors of surveying. It is difficult to raise specific design issues related to women, especially if they are seen as 'special needs', that is, non-profit-making needs, but for many women the detailed level of immediate physical practicality in housing design is the main level of concern (Saunders and Williams, 1988).

For many ordinary women 'a home of your own', especially a traditional low-rise detached owner-occupied one with a garden, seems the ideal in our society, but many men despise this manifestation of the domestic realm. But 'if it's what women want, who are the planners to tell them they are wrong?', as a young woman planner declared to me once with disarming simplicity. Some women would see the house not as a building but as an extension of their bodies: a house (and its ownership) may have many different cultural and political meanings for women as against men (Chapman, 1981: 11). Many women professionals seem to start from a completely different vantage point, and although they 'learn' the 'right' arguments from the men, they may never completely make them their own, 'they simply don't understand what we are talking about'.

A woman geographer, who was involved in research within the world of surveying, told me she was astonished at how dismissive both housing managers and general surveyors were of 'space'. Several women have commented that some of the most 'humiliating' and sexist comments aimed at women professionals come when they seek to raise issues of housing design and/or policy meeting the needs of women, whilst discussions of commercial property seldom provoke sexist comments. Obviously men do consider 'housing' to be important **really**, or they wouldn't react so strongly to women who seek to alter the city of man in this respect.

An apparently anti-spatial world view within certain subuniverses of the surveying subculture is given 'space' because one can only have one's argument against 'space' taken notice of, and 'valued', in an environment in which one's adversaries are also strongly tuned into 'space' as an issue. This is just another 'game' by people who really think alike, and therefore it is not surprising that one will find alliances developing between the spatial and aspatial (even anti-spatial) subuniverses within the world of surveying (echoes of Stewart, 1981). But both these groups are also united by a disregard of women and their needs, which are neither space nor society. Women lose out of the pro- and anti-spatial side of the debate, and both the capitalist and socialist sides of the landed professions diminish women and enforce patriarchial values.

I will now look at other land uses and consider the implications of the surveyor's world view for women, drawing on comments from students, observations from practice, and references from professional journals, the majority of which are from males, who, after all, are the majority. I phoned and talked to a range of key men surveyors involved in the planning and development of areas as diverse as out-of-town shopping centres, science parks, office development, urban renewal, and residential development. When I talked to women surveyors out in practice, I always asked whether they, or women in general, would have an alternative view of the situation.

Attitudes to different land uses

City-wide fragmentation

Before looking at individual land uses, it is important to look at the whole, as many of the problems which women encounter are the result of the fragmented view of space held by many surveyors. As stated, the tendency for a surveyor to deal with a particular site in isolation, usually from a purely commercial perspective, can create many problems for women, rendering 'social' non-monetary factors invisible. This tunnel vision, of not seeing one land use location in relation to another (or worse still seeing relationships between land uses from a male perspective based on male travel patterns), may create major accessibility and transportation problems for women.

Likewise, the tendency of surveyors to specialise (of necessity) within one specific area of practice or one particular stage of the development process, without too much reference to the implications of one action on another, further fragments 'space'. Unlike factory workers, surveyors actually seem happy with their 'alienation' from the end product! However, planners can be guilty of this too and it is not necessarily related to having a commercial (or capitalistic) world view. I can never switch off doing ethnography and was alarmed at a certain 'Women and Planning' conference to find that subconsciously my mind was telling me that I was observing a similar 'fragmentation' and concentration on detailed isolated issues without reference to the wider spatial and societal context. This may simply be the result of women trying to deal with immediate practical problems in particular locations, which is not necessarily a bad thing. However, I began to feel that the 'macro' city-wide level was being lost, and the major urban structural issues were being left unquestioned (as they are by their male counterparts too). In fact such women might well have experienced the 'same' professional socialisation process as the men in the same colleges and have been taught the same fragmented world view. In one case it was outworking itself in the commercial property world and in the other within the spatial feminist world.

Traditional surveyors like to believe in the 'goodness' of the status quo and like natural explanations that justify its existence. Urban areas only exist because people have a reason to live and to work there, this being partly the result of 'natural' factors, human endeavour, and, in more recent times, direct state intervention. Surveyors as students have great enthusiasm for geographical determinism. I get quite worried when students tell me a brand new housing estate is located in a particular site, because 'it has a good water supply, being a spring line settlement', when in reality it is because some developer spotted that

Effects and implications

that site was available cheap. Nowadays, with national grids of electricity, gas, and water, one can in theory build anything anywhere provided the money is available. It is quite difficult to suggest in this setting that particular developments, which the students believe are 'right' (because they are legitimated by 'natural physical factors') might be wrong for some people, especially women. 'Interfering' in the 'natural' development process through town planning is seen as far less natural than actively changing the settlement pattern by the activities of the market. Students have commented, 'surveyors are more land-oriented, planners are more process oriented' (for 'land' read 'nature') and even 'green belts are too social' (for 'social' read 'unnatural'). Of course women's place is 'natural' too, and so there is little enthusiasm for the idea that the system (patriarchy) itself might be 'wrong'.

Zoning and planning

Paradoxically, surveyors, in the abstract, have been strongly in favour of zoning and a macro level 'controlling view' as it is in their blood from their map-making days, but they may find that this conflicts with protecting the interests of their client regarding a particular site, and the spirit of free enterprise. They seem to be in favour of planning 'as long as it prevents someone else from developing and doesn't affect my site'. Some surveyors give the impression that they see town planners as 'absolute fools' and welcome state intervention if it creates a land shortage which can only enhance the value of their clients' land. Surveyors 'use' the planning system as another factor that 'has to be taken into account' and do not necessarily share its aims.

Planning is in a sense 'neutral' and one should not assume that those surveyors who are planners are necessarily motivated by a desire to reshape the built environment in favour of those 'less fortunate than ourselves'. Even those that are genuine in their concern may perceive the working class as 'male' and assume that they want the same things out of life as they themselves want. Within this ethos there is little place for surveyors to entertain an alternative or radical world view, based on a completely different set of organising principles such as feminism might provide.

As was seen in the historical section, the layout of our towns and cities has been traditionally geared to a male way of life in which residential areas were rigidly divided from areas of employment such as industrial and commercial areas. The latter are nowadays the areas of the highest land values, and the main arena of the work of the modern surveyor. Isn't it amazing that men consider their work so important as to create special areas for it, such as the central business district, whose prominence is enforced by the size, height, and value of the commercial buildings within it? Indeed the whole 'objective' reality of the land value system upon which the property world rests, is based on this subjective patriarchal value judgement that men's work is important. Surveyors' professional activities substantially contribute to maintaining this situation.

There are trends in the opposite direction away from traditional dispersed land use patterns which surveyors and other urban professionals have 'helped along' (if only by moving house themselves), namely the gentrification of the inner city, and the 'yuppification' of the London Docklands (Greed, 1988). Although this has undoubtedly improved the situation for professional women and men, it has worked against the class interests of many working women and men. Many 'ordinary' people are very resentful that policies that were orginally intended to benefit the deprived areas and members of society under the 1979 Inner Areas Act, have in fact benefited the affluent – 'whatever they do, they always win' – and displaced indigenous communities, exacerbating the housing crisis. The original residents of the inner city are obscured in this process.

Transportation

Women's journeys have either been ignored or included in the category of leisure, social, or non-essential journeys (Stimpson *et al.*, 1981; Pickup, 1984). As women seek to integrate outside work and domestic responsibilities, the spatial relationships and distances between different land uses become primary constraints on a woman's ability to 'cope' and achieve all her tasks, (Markusen, 1981). The straightforward male 'journey to work' is not as applicable to women. Women's journeys are broken several times combining several functions, for example, to school > to work > to school > to shops >to child minder > to home, etc. Less than 20 per cent of women have the use of a car in the daytime, as the metal beast is usually taken off by the 'breadwinner' and left to graze all day in some distant central area car-park where no other member of the family can reach it. Indeed, many men, including surveyors, get the benefit of company cars as perks (Potter, 1986) which, in the final analysis, women pay for through their shopping expenditure. Surveyors have tended to hold the same views as planners on transportation, reflecting the principles of patriarchy as embodied in the property development fraternity of which they are all a part.

Not only does this gender-blind approach adversely influence policy-making in the public sector; hard-headed private sector developers may actually have lost money because of the implicit sexism of their 'objective, scientific' approach to policy-making too. For example, too many estimates of the viability of new shopping centres, especially

Effects and implications

out-of-town ones, are based on the projections from non-gender-specific mathematical models related to 'the household', or more likely 'the head of the household', concerning levels of car ownership, travellable distances, and disposable income in the potential catchment area, for example, the needs of 'the shopper' with 'his car'. Some surveyors are still using such models even when many town planners have questioned their accuracy. There have been several examples of developers being quite surprised when their shoppers do not turn up and the scheme proves a flop. In spite of all their sophisticated computerised calculations, they have overlooked the most obvious fact that most shoppers are women who cannot drive or do not have daytime access to a car.

When I raise this issue with men surveyors, many seem unable to grasp the macro-level implications of this bias, and they come out with statements such as, 'what do you expect us to do instead – provide two garages instead of one in every house?' Compare these attitudes with the findings of an ethnographic study of a supermarket car-park (in Rochdale) which shows the qualitative realities of 'how' women use such facilities (Stanley, 1987) – with difficulty. Indeed, one senior male surveyor explained to me that when he went to visit one of 'his' schemes on a Saturday, 'I was surprised at all these men wandering around looking bored; their wives had dragged them out so they could do their shopping by car'.

Do women surveyors have a different view of transportation? Many women surveyors have commented to me, 'oh, doesn't everyone have a car nowadays' (compare with the true situation stated above, GLC, 1985). Modern surveyors express many inbuilt assumptions about the importance of the motor car that have been current since the transportation euphoria of the 1960s (*Chartered Surveyor*, April 1960, Vol. 92, No. 10: 536 and 538, which contains classic comments about pedestrians slowing down the traffic if they are not sent down underpasses to cross the road, etc.). Nowadays there are fewer special articles on the need for planning for the motor car because the car is accepted as an inevitable part of modern urban living. A whole subculture of planning for the motor car exists which spans across both surveyor and planner, and male and female surveyor. This is partly the result of professional socialisation and also the effect of the socio-economic groups from which surveyors derive. One woman student wrote that, 'no greater significance should be given to whether a woman has a car or not than whether she was wearing a red dress or a blue one'. Likewise, a male student made the not untypical comment, 'Public transport – this is a minor issue, not very important'. On several occasions in a tutorial on transportation planning, the students themselves have decided to count up the numbers of cars owned by their families, and it always comes out in the region of at least three or four per family.

Community uses

A perfect home without the provision of essential infrastructure, community provision, and social amenities is only half-way there. This is very noticeable in much modern residential development which consists of houses and nothing else. Whilst planners and surveyors are keen to meet the social and community needs of people like themselves that they can understand, they may quickly dismiss the needs of women. By definition 'leisure' is male (Deem, 1987), and women's arguably more passive needs are barely considered. If they are, they are considered of secondary importance, it can take years to get a dance centre or a creche, but squash courts and all-weather pitches spring up instantaneously. Vast areas of plans are shaded green to designate 'public' open space, playing-fields and sports facilities. Space does 'matter'; for example, golf courses take up immense areas and are used by a very small and mainly male middle-class minority. This may seem a kill-joy attitude, and perhaps I am over-reacting to the constant emphasis on sport from students, which is ultimately imprinted on the built environment in the grown-up world of the professions. For example, if they are asked to describe the land uses of an area, they will inevitably start with open space and recreational use as the most obvious and natural priority. Only after that will they move on to the most profitable uses. 'Women's' needs come very far down the list.

As indicated earlier, men can justify this expenditure as 'social' and therefore worthy, even in times of government cut-backs, as sport is seen as a panacea for all (male) social problems especially those of the inner city and 'youth' (note how 'social' can change its meaning in this chapter). Many women have been put off sport for life at school and have very little idea of what personal leisure is. However, there may also be strong support for sport as a valid cause of public expenditure from some of the women within the landed professions, for to be the 'right type' to survive in the sport-loving professional subculture, women need to be genuinely fond of sport themselves, or at least willing to cheer and support the men. These attitudes of sympathy and enthusiasm for sport and sportsmen may in fact make them out of step with the majority of ordinary women in the population, and make them the least suitable women to deal with the needs of other women.

Women working on such recreational schemes may not even think about the needs of women. A woman surveyor involved in a leisure scheme related to facilities for sailing and water-skiing said it never crossed her mind. Another woman said that such schemes were based on market research, commenting 'They employ women, don't they, in those organisations so there must be a woman's point of view included within it?' Another seemed to vaguely assume it was not the surveyors

Effects and implications

or the developers that made the decision to develop, but the planners. Indeed, in such areas it may be true that the planners are in complete agreement with the private sector, after all they are all men, 'women aren't very interested in sport in any case, so there is no point in doing anything special for them' (although resources may nevertheless be put into sport that might better be put into more socially useful facilities for women). Several women have commented to me that there is a 'boat-owning fraternity' that is strongly represented by both town planners and surveyors who have been the prime movers in revitalising the docklands and creating marinas 'everywhere'. Only 2 per cent of households own a boat, and potential waterside areas could be used for the greater benefit of the general public as a whole.

Retail development

Whereas to women, 'going shopping' is a necessity in order to get food and the other requirements of life (Bowlby, 1989: 66), to surveyors 'shopping' is transmuted into 'retail development' (Bowlby, 1988) which is an area of professional practice dominated by men where the whole aim is to make money, 'to get a reasonable return for the client'. They may bring with them into their professional work all sorts of stereotypes and false assumptions at the personal level. Many men see shopping as 'fun' and 'leisure' (because that is how they have experienced it perhaps, mooching around the shops with not a care in the world), and have little idea of the tight time schedule many women operate under and what a chore it is. Such attitudes are deeply ingrained in the land use professions. The famous Lewis Mumford (1965), in his epic work on the development of the city in history, made much of the importance of man the noble hunter and food-gatherer at the dawn of history (a book that is still referred to on many planning courses). The same Mumford, in another setting had said, 'the daily marketing [shopping] is all part of the fun' (Mumford, 1930s) when commenting on women undertaking the same process of food-hunting and gathering in modern towns. It is now becoming popular to put actual leisure facilities beside out-of-town shopping centres such as the Wonderworld proposal at the Merry Hill shopping centre near Dudley (*Estates Times*, 1.5.87, No. 892: 1). However, having 'ethnographised' this centre, I must concede that many people come on the mini-buses provided in conjunction with the development. But why should people need to travel out of town in the first place to go shopping? It's all money and time and adds to traffic congestion. In contrast, I came across several women surveyors who were from a small-scale entrepreneurial background and sympathised strongly with both small shopkeepers and the customers that used their shops.

The influence on what is built

One woman surveyor directly involved in the management of an out-of-town scheme, explained that she herself would not be seen dead shopping there, but seemed to have faith that obviously there was a demand from women in society for such facilities or they would not build them. She found it quite difficult to know where to shop herself, her office being located away from any food shops, and 'it doesn't look right to go shopping in my business suit'. She seemed to be planning for that 'abstract woman' that the men had convinced her existed. I myself had gone through a phase whilst working in a planning office of feeling uneasy about many of the plans proposed, but feeling, 'they must be right', and always returning to believing 'it's not for people like me' and almost feeling ashamed that I was so ignorant of the needs of other women who must obviously be either 'higher', 'lower', or different from the people I knew (when in reality the men were wrong). This is a common experience that women have both at college and in practice; when they cannot 'identify' the client group from their own life experience, they try to 'compensate' in order to believe the men are right.

I was very taken by a photograph that appeared on the front of *The Times* (Tuesday, 23.2.88, Part 2: 23) which showed six shopping tycoons standing outside their new superstore each holding a metal trolley (the hidden hands behind the social construction of the housewife). Compare this with the article 'Hell in the high street: your views on Britain's shops', (*Options*, April 1988: 192, Carlton Magazines, London) which demonstrated that although most shoppers are women, their needs are barely taken into account. One senior male retail development surveyor told me, 'if you provide seats you will attract tramps. Why should women want to sit down? I'm always too busy myself to sit, they must have time to waste'. Of course floorspace is money. Even if men and women do receive equal floorspace for public lavatories, the women, who on average constitute 80 per cent of shoppers, will be allocated three cubicles (one of which doubles as the disabled toilet and is quite likely to be locked), whilst the fewer men will benefit from more 'communal' provision, 'no, we are not in favour of providing public toilets, they always create a queue and that detracts from the quality of the development'. (Compare 'Courting future trends', *Estates Times*, 19.2.88, No. 932: 13.) These facilities may in reality be more important than car-parks in attracting women.

However, at certain levels of the market, retailers are purposely wooing women with children. Those developers that are currently targeting women shoppers tend to go for higher spending middle-class ones. They will change their tune and alter all the 'reasons' according to the socio-economic group involved. One such man said to me that it is impossible for housewives to go shopping with children without a car! Previously, another man had tried to convince me that even

Effects and implications

working-class women could quite easily get to an out-of-town centre by bus, if they really tried. The story varies depending on whom they are trying to convince, but they seldom ask the women themselves. The division between car users and non-car users is, relatively speaking, as great in determining how people are perceived and treated, as the male/female division (compare *Estates Times*, 22.5.87, No. 895: 12–13 'The new baby of the retail world'). Alternatively, look at just about any week of the *Chartered Surveyor* for a range of examples.

'Special' needs

Many of the details of life and land use that women see as essential have until relatively recently scarcely figured on the agenda of the landed professions. As stated, many women want decent public conveniences, baby changing areas, and sitting areas. 'Of course they don't use them themselves so they don't realise.' In many cases women's needs do not fit into the existing classifications of land use and development as embodied in planning law and development plans; creches, for example, do not fall into any specific 'Use Class' (LPAS, 1986a and b). Some feminist planners see the new 'B1' 'Business' Use Class (1987 Use Classes Order) as having 'potential' but so far test cases have met with little success. A major additional problem is that even if planners are willing to include these uses someone has got to pay for their maintenance and management. Local authorities often cannot afford to pay for these 'special' uses. Some of the so-called 'loony left' local authorities in London, where the feminist influence is strongest (particularly since the 'feminist fallout' following the demise of the GLC) have made full use of negotiation, planning gain, Section 52 Agreements (LPAS, 1986b), and integral 'condition of the permission' undertakings (DOE Circular, 1985/1, 'The use of conditions in planning permissions') to get women's policies and facilities (such as creches in new shopping developments) implemented as part and parcel of new development schemes. I have heard both sides of the argument, men surveyors cursing feminist planners and urban feminists enthusing over such schemes, and frankly the world view of the two groups is miles apart.

All this raises moral questions as to the means of bringing about feminist ends. But the world of property development (as against the academic world of theory and good intentions) is not 'pure' but is all to do with money and greed and politics and influence, and is full of tough powerful men (plus a very few equally tough women property developers too). However, when retail developers see that such facilities increase their turnover, when they go on their next development they may actually insist to the planners that they **must** have a creche in their proposed scheme. Once men see there is money to be made from implementing 'feminist' ideas, they see the sense of it all (Fitch, 1985) – developers' feminism in fact. 'It's all a matter of training your developers', an over-optimistic view perhaps, and not applicable where there is no financial reason for men implementing 'feminist' policy – that is, in most of the rest of women's sphere.

Employment

When one looks at land uses related to employment, and more broadly at the level of regional economic planning, one is struck by the invisibility of women (see Chapter IV of WGSG, 1984), and also the traditional emphasis on the importance of a somewhat 'mythical' male working class which is inevitably northern and deprived. Surveyors seem to be as guilty of this as town planners and they simply do not 'see' the women all around them in offices as 'workers'. The vast majority of workers in offices are women (Crompton and Jones, 1984) indeed they form 60 per cent of the workers in any central area, but they and their journeys to work are relatively invisible in many planning reports. In the realms of public policy-making, there are inevitable conflicts in councils between the demands of the traditional Labour movement, and the new left and feminist movements regarding the 'purity' of office development which has been associated with wicked property development and capitalism. Office development, however, is usually welcomed by women as a major source of employment although many men do not see this as creating 'real jobs' (GLC, 1984). In the commercial side of the land use professions, the question of who works in the buildings, whether they are male or female and what they think of it, are all secondary, 'it's all bums on seats' as one very senior woman surveyor put it. Their primary concern is to see office development as 'investment' for the various financial institutions, pension funds, and property companies that speculate in offices, 'property is a commodity like any other'. Office blocks are often worth more empty than occupied! Commercial development can take on a life of its own, rather like the stock market, and people seem to get smitten by pure euphoria and forget what it's all 'for'.

Vast numbers of women work in factories, but women workers have frequently been excluded from the official figures and from the image of the working class in sociological literature. In comparison, in the private sector look at any promotional brochures for new industrial or office development and you will see pictures of male young executives and happy male workers with hardly a woman in sight. Again, both capitalist and socialist men seem to be part of the same patriarchal culture. The post-war trend in town planning towards both rezoning and decentralisation of industry on to green field sites created major problems for working women. This trend continues today as industry seeks to locate near motorway intersections on the edges of urban areas. Modern high-tech science parks are springing up, for example, along the M4 motorway corridor in locations that are quite un-get-at-able without a car. These developments can be quite desolate and inaccessible for many women and are miles from any shops. However, they may be 'lucky' and have a new out-of-town shopping centre in the vicinity as is the case in Bristol, where many of the women workers on the Aztec West 'science park' shop at a nearby hypermarket. This convenient relationship between the two land uses is the unexpected result of other planning policies rather than any conscious attempt to help women.

The surveyor's influence in the development process

Having looked at the various land use components, I will now discuss the process whereby development is brought forth, and will highlight the surveyor's role in the process. Many surveyors see themselves as purely estate managers, rather than policy-makers, although the demands of their commercial world view inevitably affect the built environment. As stated earlier, the gradual removal of town planning into a separate profession has freed surveyors from having to have any concern about 'social issues'. The modern development surveyor is much more concerned with commercial town planning, that is, 'getting the best return from the site' than with socially motivated town planning which is left to local authority town planners. However, there is an overlap as many local authorities employ private sector planners from the large surveying firms on a consultancy basis, as may be gleaned from their advertisements in the RTPI journal, *The Planner* (back cover of September 1988 issue, for example) and joint RICS/RTPI get-togethers (*The Planner*, March 1989, Vol. 75, No. 3: 9–12, 'Institute anniversary dinner'). An objective land use planning ethos continues to pervade much of the work of such planners and human need is marginalised as a 'social' issue. Therefore surveyors do influence what is built in a negative sense (although one has to look for what is not there, and to consider what might be there instead to come to this conclusion).

Whatever part of the RICS spectrum surveyors were located in, or indeed whatever part of the wider 'property fraternity' other landed professionals represented, it did seem that they were not that far apart in their subcultural values as to what development they saw as being important, although they might differ as to who should own it, or how it should be designed or where it should be located. In particular, the love of sport and the prioritisation of leisure and recreational land uses was found throughout the fraternity. On the question of 'how' it happens, and whether surveyors are the instigators or enablers of development can only be answered by immersing oneself in the property world and watching what is going on. It seems to me that surveyors are part of a larger team (not one opposing team amongst several), all of whom play their part in the property development process, and all of whom are linked in terms of shared interests, social contacts, and shared ambience. Sometimes I felt that the different 'factions', and professional and ideological interest groups, were not 'enemies' but, rather, I was watching a carefully orchestrated 'game' in which they all played their traditional part (echoes of Ambrose, 1986; and of Bassett and Short, 1980 perhaps). They appeared to quarrel or negotiate, but they were all part of the same process.

Different property interest groups and professions have to have different attributes and viewpoints, in the same way that in rugby you have to have different men with different attributes, such as 'heavy' props, and scrum-halves as well as 'light' fast wings and weavers, on different parts of the field in order to play the game. Men can be different, the surveying subculture allows for both 'rough' and 'smooth' men, but when they 'play', although their differences are an essential to the game, they all 'deal' with each other in the same impersonal professional manner (even if at a personal level they have little in common, and may dislike or like each other – which confuses many women). It was almost as if they have to play out their disagreements and conflicts as an essential part of the process of 'creating' something from nothing (echoes of Marx and Hegel regarding the dialectic of 'the phenomenon of a process taking place', Marx, 1981: 254). Somewhere along the line this process takes on a life of its own and things begin to happen, often with little said openly in meetings, but suddenly one finds 'they've decided'. Regarding the model, surveyors can be producing, reproducing, and/or transmitting patriarchal property 'messages' at the same time in various directions, doing their part towards the development process, all in the same 'match'.

It would seem that the concept of 'market demand' fuses together in the surveyor's mind a mixture of 'what the people need' with 'what the investors want'. Vertically on the model, the surveyor perceives the client as being 'the man in the street' at the micro level, but at the meso level the pension fund, developer, and investor with the money to back the development are the real clients. Although the surveyor might have acquired the values of the landed interests and classes (from the macro level) he is not dealing directly with them personally in the commercial development process; rather, he (and it is usually 'he' at this level) is dealing with 'their money as the client', as represented by other professionals like himself (at the meso level) who also belong to the great fraternity. This further enforces the impersonality of the process (and the alienation from the end product), precluding the consideration of 'social' issues within this atmosphere of high finance and 'serious business'.

Many women feel quite intimidated by all this and are happy to keep in the background. Some women are involved in the euphoria of all this activity, but they seem to be more part of the transmission apparatus (as links) than actual producers and instigators of the process, in their 'helpmeet' role as research assistants, negotiators, and 'managers', and indeed as lecturers (echoes of Bernstein, 1975). It should also be remembered that routine decision-making in the areas of practice that at face value seemed least to do with spatial policy-making, such as valuation, property management, and dealing with legal contracts, were, in fact, all suffused with an internalised, but unexpressed, set of priorities which made their respective contributions to the reproduction of social relations over space.

Women and development

Different policies?

On the question of whether women have different attitudes or would develop different policies, I tended to get negative replies or comments such as 'we've never thought of that before'. However, several women surveyors have said they have been instrumental in getting more 'ladies' public conveniences in shopping centres. More alarmingly and not wishing to sensationalise the issue, a woman surveyor responsible for the management of a large enclosed shopping centre, which will remain nameless, had been pressing for better lighting, but it was considered an unnecessary extravagance. She was no overt feminist but very aware of the design dangers of the scheme to women shoppers and she had put forward her arguments in the way men understand by stressing the problems of the site layout. She stressed that more women would use the centre and spend more money and therefore there would be more profit if it were safer, but nobody listened. Then a woman shopper was murdered in the underground car-park. The developers, at last, saw she was talking sense, realising that murders are bad for business and affect the financial return on retail investment, and she was given authority to make the improvements she wanted.

However, surveyors and developers still tend to look to the town planners and the local authority to impose design conditions through the planning permission to cover 'these problems'. This is not because surveyors are always unaware of the need; however, if the suggestions for better lighting or ramps or more loos come from the surveyors, then their clients (usually large financial institutions) may accuse them of being extravagant with their money or veering towards the loony left. The reality is that people will not use their developments if they are too badly designed. Therefore, to save face and not to appear uncommercial the surveyors 'need' to have the planners take the role of the socially concerned and to 'force' them to implement such improvements (another part of the development 'game'). I suspect the subtlety of this negotiating 'game' may be lost on some women surveyors who do not compartmentalise their lives or their professional practice into 'commercial' as against 'social' but see it all as 'practical' and 'obvious'.

Unfortunately, many town planners are even less aware of women's issues than surveyors and so they are not going to write social and safety requirements into their planning statements in the design and negotiating stage. Nor are they going to be particularly keen to enforce them if it puts the developers off developing altogether. Even today many town planners and chief planning officers are surveyors with a commercial or building background rather than a social or modern planning background.

Women and property

A theme that struck me throughout collecting material for this chapter is that whilst many men are so confident that 'what is, is' and that the status quo is right, several women manifested an ongoing inner conflict as to what was 'right'. However, with time, many women learnt to suppress this uneasiness. It would seem that some women have 'learnt' how to develop the perfect impersonal professional demeanour, never thinking about themselves or their needs, but acting as the impartial agents of those they serve. Indeed, many women have been told all their lives that they must never think about themselves and that they are 'selfish' if they do so (selfishness is the cardinal sin for women according to Gilligan, 1982). If they always think about the needs of others first, it becomes only 'natural' for them to take on board all the values of the surveying subculture as the 'other', even when it is against their better judgement. Perhaps this explains in part why many women surveyors were not radicalised by their involvement in the property world. It simply had not occurred to them to see what was happening around them, which they were involved in promoting, as being 'for them'. Such are the powers of professional socialisation that they have been blinded from seeing the implications of such polices for their personal lives as consumers of urban goods and services.

I came across several women who were responsible for vast sums of money, and who knew everything there was to know about the property market, but if I asked them about their own property interests they would appear surprisingly *gauche* and would make the traditional reply that they left that sort of thing to their husband, or they had never really thought about it. There seemed to be an absolute division between their professional and private lives. Some women and men commented that this was partly because people wanted to retain a sense of 'romance' within the haven of the home which would be undermined if the same approach to legal and financial matters used in professional practice were applied to domestic arrangements. However, several older women had subsequently 'got wise' as a result of divorce and personal financial disaster; and it would seem many young women surveyors are far less embarrassed about discussing and dealing with their own salary, taxation, investment, and pension arrangements. Those that were concerned about money, including those that had become business women, could not be seen as incipient capitalists by any stretch of the imagination, and did not appear to be motivated by a desire to exert power over others, or to be 'rich'. Rather they saw involvement in the business world as a way of gaining 'freedom', or had got into it 'out of force of circumstances' because of the need to support their children or make provision for other 'future difficulties', having found the employment opportunities offered to them by existing organisations 'too inflexible' or 'difficult to manage with all my other commitments' (echoes both of Hinchcliffe, 1988, and Joseph, 1980). In fact many were paddling their own canoe – not for themselves, but for the benefit of others.

Some such women felt doubly 'judged', as they were made to feel they were selfish or disloyal in taking this option, in becoming independent or (temporarily) 'rich', but at the same time they knew they would be condemned for their imprudence if they didn't. 'They condemn you for making money, but the state doesn't provide any alternative for women, all they want to do is tax your own hard-earned money as investment income, you can't win.' It would seem that in the present scheme of things whatever women do they are seen as 'wrong', often being attributed with motives and values that would be more appropriate to powerful men. Indeed, many women who wanted to set up on their own in property development had immense difficulty getting any credit from banks and other financial institutions; even the domestic mortgage market is unequal (Nationwide, 1986). In a world where one needs money as well as good ideas to shape the built environment through initiating development, not being 'trusted' by the 'old boy network' forms a major barrier. Such problems affected both women who were seeking to set up 'feminist' housing ventures and women who sought to emulate male developers. You have to have money, or inspire confidence that you are 'the right type', in order to get future financial backing if you want to directly shape 'what is built' through building it yourself. Although women may influence policy within local government town planning 'a little bit', they have hardly got anywhere as property initiators within the private sector in their own right.

Many men appeared to believe that it was not right for women to own anything, particularly at a time of housing crisis, whilst the men generally assume their right to at least one house (and wife to go with it). Women surveyors were frequently 'reminded' of their selfishness and blamed for the problems of the world, 'you don't need anything, it's not for people like you' or 'you should think about the working class'. Indeed, this supposed concern for 'the working class' (albeit expressed as an abstract concept) seemed to be used by some men surveyors, particularly in local government, as a talisman to legitimate their right to professional status (and often to justify their resentment of middle-class and professional women). Many of the women considered they did 'need' money, especially if they had to 'paddle their own canoe' against the rapids of life.

Of course the extent to which women should be involved in 'the system', be it defined as patriarchal, capitalist, or whatever, is a major ideological issue. Indeed there is a potential conflict within 'feminism' itself as to whether women should seek 'equality' through personal advancement on an individual basis within the existing 'system' (and change it in the process), or through remaining separate from it and seeking to challenge and change it from without. Whilst some would argue that the former route only serves to enforce class inequalities between women (and men), others would argue that the latter is almost impossible to achieve, 'after all, we're all joined to the same water and sewerage system, we are part of it – if we like it or not – and we're not in a powerful enough position to go it alone, you can't be a surveyor in a vacuum without having the business world and male clients'. No doubt each reader has her own way of resolving these dilemmas.

Conclusion

It would seem that whilst the values of the surveying subculture do affect what is built, as described above, it is not just the men surveyors who subscribe to these values, as many (but not all) of the women held a similar commercial perspective; as several women have put it, 'it's what my job is all about'. Also, the surveyors are part of a team of property professionals, including the planners, who all contribute their ostensibly 'different' (but regarding women's urban needs, often remarkably similar) viewpoints to the process, but within the enterprise culture of the 1980s the surveyors, and the financial institutions whom they represent, do appear to have a major influence on 'what is built'.

Chapter eleven

Conclusion

General observations

Many of the problems that women surveyors encountered were nothing to do with land use and development. Rather, they were matters of organisational structure, interpersonal relationships, and ethos which might be experienced in other professions too, that is, aspatial (social) rather than spatial factors related to gender. However, the particularly 'spatial' ethos of surveying with its emphasis on 'land', combined with a somewhat spurious emphasis on technology, reverberated throughout the subculture and gave an extra 'gloss' to many of the issues which women encountered. These factors acted as major constraints, for if women were perceived and treated in a way that was detrimental to their true personalities and needs; if they were 'not allowed through', or shunted into powerless 'helpmeet' roles (at the micro level), then they would be unable to exert any influence on the nature of the surveying profession (at the meso level), nor, consequently, on the nature of the built environment and society (at the macro level).

Despite the increase in the number of women entering surveying they still comprise less than 4 per cent of the qualified membership of the RICS, therefore the influence of the male majority is inevitably more significant than the effect of the small, yet dynamic, number of women in the profession. It appeared that many of the men still 'see' the world of surveying, and the inhabitants of the built environment as 'male'. They assumed a male audience, as shown by the style and content of articles in the journals, lectures in colleges, and their professional reports, and policy-making. When they 'remembered' about women and dealt with them as a 'special' topic they seemed as capable as the men in any other profession of making the right noises related to some hypothetical 'abstract woman'. If they were caught off guard, little appeared to have changed. Indeed, women's needs were often seen as a 'social' or 'design' issue whilst leisure and sports facilities would never be seen as 'special'.

Conclusion

Paradoxically, surveyors' emphasis on 'reality' and 'the practical demands of the market', reduced the likelihood of them translating their biased gender views into spatial policy, which would be seen as 'interfering'. Nevertheless, their role in perpetuating the status quo could be equally detrimental to women's interests, as many of their commercial decisions have social implications for women. In contrast, within the subuniverses of housing and town planning, the 'socially aware' or even 'socialist' surveyor or planner was potentially much more of a problem for women, as 'he' sought actively to change the society and 'help the working class' (as perceived through patriarchal eyes). Women who raised women's issues were likely to be told that they were being selfish or 'too middle class'. Such attitudes 'fazed' and confused women, diverting them from their original goal. Generally, a more condescending attitude to women at the interpersonal level was reported in these specialisms than was found in either the more technological or commercial subuniverses of surveying, where the manner was 'more businesslike', and women were more likely to be valued as an efficient 'man'power resource and encouraged.

Nevertheless, I felt a mixture of criticism and great admiration for some men surveyors! Some appeared more aware of wider environmental and social issues, such as conservation and regional inequality, than some women surveyors! One also had to admire the sheer effort and detailed work involved in their contribution (albeit somewhat misdirected) to the development process which creates the built environment and which is often taken for granted by the general public. Some of the more traditional men seemed as concerned as myself at all the changes that had befallen the profession in making it more commercial and modern. Also, in spite of the apparent existence of 'patriarchy' at the structural level of society, which operates 'normally' to the disadvantage of women, there are exceptions to the rule. There is no doubt that some individual men surveyors were helpful, enlightened, and supportive, and certain women were seen as competitive and unhelpful (many women surveyors recounted experiences in this respect and many much preferred men as working colleagues). Whilst many women surveyors have not progressed as they would have wished, there is a significant minority of women in both education and practice who now occupy senior positions.

Women (but not all) who reached positions of influence were not necessarily going to hold an alternative perspective towards professional work, or the nature of the built environment, and many were not directly involved in urban policy-making. Some were remarkably 'straight' and uncomplicated people, and tended to distance themselves from more feminist colleagues (although, arguably, their careers had benefited from the changes brought about by the women's movement).

Those women who were conscious that all was not fair tended to concentrate, quite justifiably, on the aspatial (social) issues affecting their careers rather than spatial issues which affect all women as, 'you can't do anything about it in any case'. Whilst one must change the aspatial factors to create the right conditions to enable women to be taken seriously in order to change the spatial situation, to simply change them as an end in itself so that more women might 'succeed' at an individual level achieves little for women in society as a whole. Indeed, the nature of the surveying subculture remains undisturbed, whilst ostensibly giving the impression that the profession is progressive because it admits women and enables them to succeed. However, many would argue that there is need for more fundamental change: indeed, a complete reformulation of the surveying profession in order to take into account fully the 'different' needs of women as fellow professionals and consumers of the built environment.

The position of women in surveying

Although women are now admitted to surveying, in general their role is still ancillary rather than central, as 'helpmeets', although they contribute a great deal to maintaining the surveying subculture (albeit against their better interests). After the initial build-up (the 'splutter effect') and a certain hostility, men have learnt how to 'deal' with women and use them to their advantage. In education, for example, women students may sometimes be admitted to surveying education, to help 'protect' male interests in times of potential 'falling rolls' and cutbacks; and at other times 'equally' on their own merit. Some men, no doubt, find a conflict between dealing with women surveyors as part of the professional 'tribe', and treating them as women. This seems to be resolved by creating a hierarchy of 'value' in which women, whilst being welcomed, are often perceived differently from the malestream students. This is not necessarily linked to attainment or intelligence. 'I just don't know what they are looking for, it doesn't make sense'. I never did quite 'crack' what Factor X was. Obviously some women possessed it, as evidenced by their acceptance and success.

In surveying education, there was little evidence of outright conflict and opposition as described in some of the literature on class and gender; indeed, the women were often welcomed and encouraged by the men. Rather there was an overall feeling again and again of 'drowning in a blancmange' or 'fighting in a fog'. Education is a 'deviant' activity in the surveying subculture, and, as stated, less than 1 per cent of all qualified surveyors (2 per cent of the women) are involved in it. Doing well in education is not the name of the game. Also, women (and men) have already undergone substantial 'filtering' at entry and during

Conclusion

surveying education, the roots of which go back into the school and family background of the applicant.

On entry to the world of professional practice, and subsequent progress within it, 'closure' on the basis of 'class' and being 'the right type' plays a central part, in addition to gender, in deciding who eventually ends up where. Whilst 'background' can sometimes explain why some women succeed and others fail, 'class' is not everything as one must be totally acceptable and 'tactful'. Even women who would not see themselves as feminist appeared to be instinctively treading warily, making allowances, and generally over-compensating for men's 'attitudes'. Men can 'sense' when a woman holds an alternative world view, and so high status women who hold 'feminist' views may in fact be less acceptable than lower status, but more conformist women. Some women appear to be 'acceptable' when they are really more radical and enlightened than they look; whereas others simply hold two world views at once without realising the inherent contradictions, and may simply want to 'please' and do 'what is right'. Nowadays, some younger women surveyors, albeit non-feminist, have undoubtedly imbibed many of the ideas of popularist feminism from the wider cultural setting of the 1980s and use its lessons and strategies in developing their careers, expressing their 'natural' right to equality in a quite unselfconscious manner which often astonishes older, more cautious women surveyors.

A major sensitising concept was the fragmentation and isolation of individual women within surveying practice. They might be, variously, geographically separated from each other, expected to keep their status distance from other women in the office, ill at ease with 'ordinary' women in the community, and alienated from others of their gender as they sought 'to be the same and fit in' (with the men). Likewise, women's needs, whether spatial or aspatial, were marginalised or made a 'separate' or 'special' issue within the discourse of surveying. Parallel to this, the fragmentation of the surveyors' world view, in dividing up the built environment into separate land uses and professional practice into different specialisms covering different stages and aspects of the development process further, made it difficult to grasp and present an alternative 'total' feminist world view which related to the whole built environment.

I was struck by the sheer 'caprice' and unpredictability of the situation. You never knew how men surveyors would 'treat' a woman in a particular instance. The situation has not yet reached a point of equilibrium following the influx of increased numbers of women, and men surveyors are still trying out different 'solutions'. Part of the 'problem' is that whilst men know that men come in certain standard types, and they can always find the right sort of chap to fit the right mould, women are much more of a risk as they do not 'fit'. Men haven't

Effects and implications

yet sussed out the range of 'types' of women and how to treat them ('grading' women according to body shape is of little value when assessing their aptitude in professional work; but several women have commented that all the women, chosen by certain male gatekeepers, look remarkably similar). Women surveyors come in a greater range of 'types' than the men, which is not surprising as one had to be a 'one-off' (particularly in the past) to choose surveying in the first place. Whilst some 'left-wing' men are of the opinion that most women surveyors are either 'sloanes or Thatcher clones' this is very far from the truth (and derogatory). Also, some women surveyors seem so much 'brighter' and more dynamic than many of the men, and even if the women are 'only trying to be helpful' this can be very threatening.

If a woman is successful one must carefully investigate the effect of her success. Openly, feminist women may find that their role is to act as a 'contrast', to be the Aunt Sally, thus unwittingly reinforcing the values of the subculture. One must look at the long-term effects of their initial success on their future prospects. Women who are successful in male technological areas may find that their very success as a 'specialist' debars them from further progress as they have become 'indispensable'. There are certain technological areas of surveying where the men are leaving 'for something better' and women are 'taking the leftovers'. There are also many examples of women who have made an initial rapid rise, and then are moved sideways into a dead-end job, or remain stuck for years whilst being overtaken by younger men. It would seem that some men think women will be content with only limited success, or they see the employment of professional women as a temporary measure to tide them over during a 'man'power crisis. Nevertheless, a minority do succeed and can act as positive role models for others.

Implications for theory

Whilst 'gender' seemed a '100 per cent' important factor in many women's lives in affecting how they were treated in surveying, one had to admit that many of them would not be in surveying in the first place were it not for their 'class' position. In my model, whilst 'macro' level world views based on class and gender as 'first causes' were of great value as organising concepts and baseline explanations, there were many other 'secondary' factors that needed to be taken into account, such as age, personality, motivation, and personal life experience; rather like putting tracing paper overlays showing additional detail on top of the initial Ordnance Survey base map when doing a site analysis. Indeed, as women surveyors themselves pointed out to me, 'women surveyors are all different, we're all individuals, you can't generalise about us'. I had to agree that current theoretical viewpoints seemed

Conclusion

somewhat impoverished in the light of 'the tapestry of life' which I was unravelling in my research.

Human beings are amazingly complicated and whilst 'types' are of value, to say that a person is always going to be treated in a particular way, in a particular situation, because of their class and gender is an oversimplification. Sometimes women behave 'wrongly' out of self-interest, and sometimes individual men are very helpful. It was also difficult to pin down and separate out whether women were experiencing the effects of patriarchy, capitalism, or indeed racism, in a particular situation at the micro level of the model. Such a question might seem ludicrous in the heat of the moment when one is experiencing 'trouble'. One black woman surveyor explained to me, 'when they are rude to you, it's difficult to know if it's because you are black, female, or they have had a bad day'.

Sexism was not always manifested by nastiness, quite the opposite in fact. Many women pointed out that women could suffer equally when men were pleasant to them, 'paternalistic, patronising niceness' could be even more deadly in the end than obviously hostile attitudes, 'they can be friendly and still put you down'. Men can also show great inconsistency between dealing with women *en masse* and at the individual level. Large numbers of women students can be invited into surveying education at one level (without being seen as a threat), but further along individual older women out in practice may find themselves being ignored when they seek advancement, 'thus far and no further'. (In spite of 'man'power shortages, there is still only limited space at the top, because the profession is so pyramidically structured, a problem which changes in the organisational structures of surveying might improve.)

Of course, women surveyors have not suddenly sprung up from nowhere simply because of the second wave of feminism. Surveying is so much bigger than just a profession, or a particular set of areas of expertise, it is truly a tribe and a subculture, even a religion. Women have always had a role to play in surveying within families, offices, and surveying dynasties. Many of those entering surveying today are 'safe' women who are not going to question things as they do not wish to jeopardise the interests of the tribe to which they belong, or their wider class interests. Those women who enter surveying with more 'independent' or radical viewpoints may be outsiders, marginal women, or insider women who are going through a temporary phase of rebellion or doubt. The surveying subculture can 'use' all these sorts, and may even promote a marginal woman, if it serves the interests of the 'game'.

The advantages of belonging to a privileged 'class' and having a higher income than many women (albeit through hard work) could offset the effects of gender so much so that women surveyors did not necessarily share the same life experience or problems of lesser women. As one socially aware woman surveyor put it (who could objectify herself with alarming honesty), 'really I suppose I have no idea how the average housewife lives – any more than a man'. Many were not 'conscious' of what 'all the fuss was about'. In many cases they had never encountered feminist literature, so if they had 'problems' they were unlikely to frame them in terms of feminist analysis. However, some women (and men) surveyors do have an alternative caring world view, but this may have to be 'suspended' whilst they are operating as surveyors. Some had to fight quite hard to suppress it. As one woman put it, 'sometimes I get this burning feeling in the back of my neck, it's my conscience I suppose', but then, 'surveyors are not social workers', and (quite logically) 'we have got to make a profit for our clients'. Of course many women surveyors would argue that men are reducing their opportunities of making profits by ignoring women's needs and market potential.

Even when women achieved a 'class' position comparable to men, it would seem that other meso level social and cultural factors held them back, or alternatively it took time for their own personal attitudes and those of their colleagues at the micro level to catch up with their new status (echoes of 'cultural lag'). Some women 'had got their act together' more than others, and seemed to possess more 'cultural capital' to draw on in times of crisis. Regarding the question of the relative power of structure and agent in society (Giddens, 1984) (that is, which way the arrows go on my diagram), it seemed to me that women surveyors were not 'passive' recipients of deterministic societal forces, but some were better than others at 'getting by'. Both individual personality and the level of support and 'forewarning' from others determined whether or not they 'made it'. Some women seemed totally unprepared and ignorant of what they were to encounter, whereas others seemed to be living, 'as if they've been through it all once before, and so they always know what to do'; this blessed state apparently being reached because other wiser women shared their life experience with them (especially mothers and sisters), actively helped them along, and acted as 'role models'. Others simply possessed exceptional amounts of native cunning, or *savoir faire* or were even seen by other women surveyors, as 'not very nice people, too pushy by far'.

Notwithstanding the importance of personal factors, in addition to class and gender, that could modulate a woman's life experience, I was aware from my own life experience, and from what other women had told me, that in daily life people were often judged and treated on the basis of mono-dimensional stereotypes attributed to them because of their perceived class, gender, and background, 'you can tell what a person is like within the first thirty seconds'. Indeed many men surveyors seemed alarmingly prone to such generalisations in spite of the momentous changes that were meant to have occurred in society's perception of women's capabilities and social position.

From studies of macro sociological literature, I knew that women who imagined that they were operating 'independently', 'as I thought best', were really constrained in their choices, to a degree, by powerful underlying economic forces within society of which they might be totally ignorant but which nevertheless existed. Therefore I would argue that there is a need to develop theory that retains the underlying dimensions of class and gender, but which is capable of seeing how these factors are 'modulated' in an individual's life by other 'personal' factors, so one can explain why a person is as *she* is (and not like her sister), as well as giving a 'total' explanation of society itself at a macro level. Ethnographic and more open-ended approaches to research may be the way to achieve this in contrast to the more 'closed' approaches of the past, in which, for example, questions about 'class' provided little 'room' for women's experience or for 'long-winded complicated irrelevant explanations' (which I always welcomed).

Overall patriarchy seemed to be stronger than 'class' or the ideological divisions within the surveying subculture, even though there are some quite major differences between male surveyors, both horizontally in terms of different specialisms, and vertically in terms of different status groups. Within the wider world of surveying practice, it also seemed that male technicians and other ancillary people worked 'with' the male surveyors in maintaining patriarchy, including 'colonised' office women, who might be used against high status professional women (although others were very supportive).

Patriarchy also seemed to be stronger than 'professional expertise'. Several women told me of examples of men, who appeared to them to be under-qualified and inexperienced, being chosen above women who were eminently more suitable. In education, there was at least one case of a woman taking her college to an industrial tribunal. Interestingly, such problems seemed to be greater at the more 'social' end of surveying than the technological end of the spectrum (where men might 'welcome' women), no doubt because there is less of a social policy element in the latter, and not enough women to be a threat. Indeed, the definition of what counted as 'man's work' seemed to change with the changing requirements of patriarchy, both in lecturing and professional practice. For example, as was explained in the historical section, land surveying and mapping were once seen as exclusively male high status preserves, whereas nowadays they have become relatively more open to women who are likely to be employed at a quasi-professional or technician grade. With the coming of computers, certain map-making operations are now seen as a spatial version of 'word-processing' that can be done by women. However, men will still keep certain areas of 'high technology' computer application for themselves, such as their predominance in satellite geodesy, thus creating a gender hierarchy within a recently evolved professional area (FIG, 1983).

At the macro level, it did seem true that 'land' was the centre of the 'world view' of surveyors as found by Joseph originally, but I found this concept of 'land' had little substance as a concrete spatial reality. As explained in the last chapter, surveyors were ill at ease with spatial design issues. I had to go further and unpack the true meaning of 'land' which often seemed to be a euphemism for 'how the world ought to be'. It was more powerful at the subcultural level as a value, which could mean anything surveyors wanted it to be to legitimate their world view. For example, 'land' provides linkages with the surveyors' illustrious past in association with 'landed estates and interests'; it can legitimate the surveyors professional monopoly in association with 'land surveying', and it enforces the surveyors right to rule and exercise control in the sense of 'land' management (and 'estate' management). More recently it has acquired wholesome 'green' connotations. Throughout this research I was fighting my way through a hall of mirrors, of false images and 'fronts' that needed to be looked behind.

Whilst the current emphasis on 'land' management is likely to be linked to the needs of the commercial market, and in its present incarnation surveying is highly capitalist-oriented, surveyors could equally flourish under more socialist situations and still retain control over land. The entrepreneurial/bureaucratic division seemed to be, relatively speaking, two sides of the same patriarchal coin, as in the final analysis all surveyors are 'brothers' (echoes of Battersby, 1970). However, at present surveyors find entrepreneurial women the most usable and least challenging type to patriarchal power. It seemed to me that 'feudalism', as an organising concept for society, had room within it for the surveyors' love of 'control' over land and for the brands of both 'landed capitalism' and 'paternalistic socialism' that surveyors espoused. It was also eminently 'patriarchal' to provide the ideal world view for the average male surveyor (although, interestingly, some men said it was not that wonderful for them either). The whole subculture of surveying seemed to be suffused with a touch of 'cultural lag' which put them a little out of step with the rest of the twentieth century, and which gave it a rather eccentric bit of quirky unpredictability and gentlemanly chivalry which were sometimes to the advantage of women.

Whilst uncertain in my own mind, even at the end of the research, if pushed I would say that a somewhat neo-Weberian feminist interpretation equated most closely to what I was observing, as expressed by French (1985). However, I felt uneasy throughout this research in using traditional definitions of 'class' and 'status' (which were invented to explain the experience of men) for women, for fear of being misunderstood or tying myself down. I have therefore sought to describe and illustrate what is going on in the world of surveying as a means of explaining what 'makes it tick', rather than using precise definitions or adopting a final theoretical position. I have mapped out the parameters as expressed in my model, and invite the reader to make her own horizontal and vertical linkages between elements to see 'how it works'. Likewise, I have a certain uneasiness with 'reconstructing' the world of the woman surveyor, both past and present, on the basis of modern feminist 'orthodoxy' as the motivations and values of my 'subjects' simply do not fit into theories that evolved from a relatively radical academic heritage. Women surveyors might have appeared to be doing feminist things, but they were motivated by a very different set of values ranging from a 'one England' sense of social responsibility in the past to a belief in the enterprise culture today.

Hopes for the future

The context
Larger proportions of young people are going to college and there is an enthusiasm, from both males and females, for courses leading to professional qualifications such as law, accountancy, and surveying. Whilst in the past, professional education was somewhat restrictive as to entrance, it is now becoming more open, and more women and other 'outsider' students go on surveying courses (because of falling rolls of 'normal' school-leavers). As a greater proportion of the population join the professional classes, I suspect that not everyone will have the 'same' destination or 'value' as a result of participating in the 'same' educational process. It should be remembered that although over half of school pupils are female and many do 'well' at school, this has very little relationship with their distribution within the occupational structure.

Expansion and structural change is occurring out in practice and the nature of the male surveyor 'himself' is changing. Surveyors keep telling me that the days of the traditional partnership system are numbered and that in the future two-thirds of surveyors will be working in incorporated organisations rather than in firms based on the partnership system. A change in organisation, or an increase in the representation of women, should not be equated *per se* with greater 'equality'. Rather, it suggests that surveyors are regrouping and creating a different way of expressing the hierarchies and divisions within practice to accommodate a wider range of levels and types of surveyors and fields of expertise. At present they are still sorting out their strategies regarding women, and 'caprice' reigns. As the surveying tribe gets more

Effects and implications

adept at directing women into 'suitable' areas of practice, a more institutionalised and stable gender hierarchy will emerge within the profession. I have my suspicions as to how 'permanent' women's new presence in the profession is likely to be. I tried to imagine what the scenario would be like without a buoyant property market and Conservative government, and without a continuing demand for more 'man'power in the professions. (The demographic time-bomb argument is only valid whilst there is an expanding economy; in other circumstances a drop in population might actually be viewed as 'good', the planners having always told us that we all live in an 'overcrowded' set of islands.) Were this situation to arise, I suspect that men would continue to be recruited but women would be gradually 'phased out' of practice (except for the perennial 'exceptional woman'), but they would retain an important tribal role in the home, newcomers bringing it 'new blood'.

Very little has been done, either in education or practice, to accommodate the 'different' needs of women. It would seem an ideal opportunity to use current changes and 'man'power crises as an 'excuse' for raising the question of women (bearing in mind that the demands of feminism and the demands for more 'man'power for the profession are not the same; they just happen to coincide in an uneasy alliance at present). Several young women (and a few young men) surveyors, including those involved in the JO, have tried to raise these issues with the RICS itself. It is in the interests of the older men to take seriously the demands of young women surveyors (and some prominent men, to their credit, are doing so), particularly if the number of women entrants increases, as otherwise they may find that they are going to experience a 'manpower crisis' 'eating into the very fabric of the profession', similar to the one which the Law Society is currently undergoing as a result of recruiting 'too many women'.

It would seem that, in the past, men surveyors have thought in terms of meeting short-term need when recruiting women (echoes of Brett-Jones, 1978); or in terms of their having a subsidiary supporting role. It may have never even occurred to some of them that the women would want to be women (and mothers) as well as being surveyors, and possibly have children too (such attitudes can only exist in highly patriarchal societies which fragment work and home) and become partners and continue working until they are as old and grey as senior, distinguished senior men in the professions! (The whole issue of the likely future increase in the numbers of older working woman professionals has hardly been addressed, anywhere, as yet: another time bomb for the start of the new millenium.) Many young women would feel more secure and encouraged if they were convinced that the men in charge were aware of these realities and appreciated women's specific employment needs (indeed, many women are sole 'breadwinners') both in terms of the patterns of day-to-day work, and the organisation of life-long career structures. Future career patterns and employment models for surveying need to be based on accommodating the various phases and stages of women's lives so they can give, and be used, to their full potential rather than being based on the traditional, more mono-dimensional, linear male model of the life of the professional gentleman.

There seems to be a new mood among many women in the professions. Whereas in the past, women's problems used to be framed in questions such as 'how can we change the profession?', nowadays women are more likely to state, 'WE ARE the profession, let's formulate ways of reorganising ourselves and our colleagues to enable us to operate to our full capacity as both professionals and women' (after all, one definition of a professional body is that it is a mutual aid society, in existence for the benefit of its members). Already some of those women who have male 'partners' (that is husbands, etc.) who are also in surveying are expecting their menfolk to modify their work patterns and share the caring domestic role with them (I have actually come across a few men staying at home whilst their surveyor wives work). Again it should not be assumed that only the women must change, for as fewer young men surveyors can find themselves traditional wives who stay at home, and instead marry career women, the men must shoulder a greater amount of the burden too. This means it is not just a 'woman's issue' but the whole nature of the professions and the work/home division of labour must change.

Changes in education

Provision for women in college should not be seen as a 'special extra' but as a necessary and integral part of student and staff back-up. Child care and more flexible approaches to study should not just be seen as a 'woman's issue', but as changes which men can benefit from too, if only to avoid yet more falling rolls in twenty years time, or the extinction of surveying dynasties. In particular, attendance hours need to be reconsidered on a day-to-day basis. There is also a need for a move away from the solid three-year course which is 'normally' undertaken straight from school. Many women resent the fact that time spent in education and getting established out in practice in their twenties eats into their potential early child-bearing years, particularly if they want to have their children whilst they are still young.

There is a need to introduce a woman's perspective and dimension throughout the syllabi. This is virtually impossible because of the factual, technical, and peopleless nature of much of the material. The vast majority of lecturers are men, many of whom have little awareness of women's issues, and many of the student groups are still composed chiefly of males. It can be quite difficult to teach a predominantly male audience anything verging on urban feminism, unless it is done in a totally unselfconscious matter-of-fact way as if it is nothing out of the ordinary. However, if the approach is too impersonal, the danger is that students will 'learn' the material and trot out the right sentiments, and then totally forget it without ever having internalised it (as already happens to some extent now with other 'alien' material). Worse still, the Aunt Sally booby trap may be triggered if the lecturer is a woman, and everything that is said will be used to confirm to them that feminists are mad, thus reinforcing the values of the subculture. Some women lecturers favour using role play, and this does work quite well on housing students, but many male estate management students feel very awkward about this, so much so that the actual content of the role is forgotten amidst the embarrassment of doing it. Also, in the more technological areas of surveying there are no 'social' subjects which would lend themselves to this approach.

Some women would say that it is not the women lecturers, but 'the men who should be taking this on board', and deal with what is in fact a problem created by men in the first place (even if this involves some prompting to put the idea into the men's heads). Others would argue that this suggestion is playing into the hands of patriarchy, and that no man can ever really deal with these issues because he has never experienced them as women have, and he is an irredeemably biased party. In comparison, the RTPI now has a policy, which has had a mixed reception, that the needs of the disabled, women, ethnic groups, and 'other minorities' should be taken into account in teaching material. Many polytechnics and local authorities now have an equal opportunities policy which, some women argue, includes how and what people teach. However, many men seem unaware of the applicability of this to them.

Of course, if only women teach women's issues, or if it is subsumed under the special category of 'social issues', this may achieve even less. If men incorporate the question of women into their lectures they may do more harm than good in giving a false impression that all is now well 'because we've done women', or that women are such helpless crippled little creatures that women students will be (as some women have put it) 'put off from being a woman for ever after', because 'he made us squirm and feel like we were spastics, I hate being seen as disabled, with all the boys glaring at us'.

In the content of non-'social' subjects there is not much scope for feminist material but there is much scope in methods of teaching, use of examples, and general 'girl-friendliness' (Whyte *et al.*, 1985) especially in technological subjects. However, the increase of women lecturers which is noticeable at the 'junior' level across a diversity of surveying subject areas, is very heartening, and it would seem that surveying is ahead of geography and many other subjects in this respect.

Changes in practice

The whole question of provision and financing of facilities for women and child care needs looking into, especially in the private sector where increasing numbers of women surveyors are found, possibly using the proposals in the report produced by women solicitors as a basis for discussion (Law Society, 1988; Nott, 1989). There are special problems in the surveying profession, in which many of the women who wish to have children are in their thirties and at associate partner level and therefore are not covered by employee legislation. There needs to be fundamental reappraisal of the scope and nature of deeds of partnership to ascertain which of women's 'special' needs should be taken into account, and possibly paid for, within the fee-earning structure of the whole partnership (as against the income of individual women). Whereas, in the past, women were expected 'to go off and be earthy elsewhere' and have their children at a safe distance from the 'real world' of work, nowadays, increasingly, it would seem that many women surveyors want to be both mothers and surveyors simultaneously, and maintain a full work-load. Ideas such as 'career breaks' are not particularly popular, as women feel they will lose touch with what is going on if they are away for any length of time from a dynamic and rapidly expanding property world.

None of this can be looked at in isolation from the (inadequate) levels of government provision of child care, and 'caring' support in Britain. (I came across several elderly single professional women who had devoted years to caring for elderly mothers: 20 per cent of the whole population will soon be over retirement age.) Many women are of the opinion that if the state is unwilling to provide facilities for 'carers' (and some women surveyors are unhappy with the rather threatening 'socialist' image of state provision in any case) there should at least be full tax exemptions for child care costs (as applies to some extent in the USA) and possibly subsidies and incentives for more people to set up as child care providers. These needs should be met in a fully business-like manner, in the workplace (for example, in prestigious West End surveying practices perhaps or in shared schemes amongst small provincial firms in the same town), or in residential areas. Everyone seems to have such difficulty getting secretarial and clerical staff, so perhaps such facilities should be open to non-professional staff too (experiments in child care by the main high street banks are being watched with interest by many women in the professions). Ironically, if one of the aims of the profession is to serve society, it is very unfortunate that many women have no option but to put their own personal social responsibilities second, in order to hold their own in the profession.

Many women favour a more flexible approach to work itself on a day-to-day basis (as child care is an ongoing daily commitment, not one that can be blocked off into weekends, or a couple of months every so often). One of the marks of the surveying profession has always been (as Joseph explained, 1980) that men are attracted to it **because** of the opportunities for 'freedom' in organising their work, and 'not being stuck in an office all day but getting out and about'. Women would like to apply and extend this ethos to their own particular circumstances. Many women feel that they could achieve just as much per day, if they were 'given the benefit of the doubt, and trusted when out of the office' and were given the freedom to fully combine professional and domestic duties in the way most convenient to themselves. This already happens with many women in sole practice and those that sell their professional expertise (sometimes to immediate past employers) through consultancy work.

Might this 'freedom' also be extended to those who are employees or associate partners? This flexible approach involves more understanding and a spirit of trust from the people (usually men) keeping an eye on the day-to-day management of a practice, and more give and take all round. (All firms will eventually be in the same boat, so one would not be at a disadvantage in adopting this approach.) For example, women might choose to work more at home (as well as in the office), in the day and at evenings and weekends. Rather than 'work' being measured by attendance between certain hours within the office, people should be given a time-scale in which to complete an assignment and be given the freedom to decide when and where they will complete it. Some may prefer to work 150 hours one week, and ten the next (especially at times of school holidays) and be more productive as a result (this is what already happens for many men, if the truth were known, particularly when they have 'got a big job on').

Of course, since so much professional work involves working in teams and dealing with others there would be a need for considerable organisational skills here. To be business-like, this approach may require a more formalised 'time-keeping' for both men and women, as is already common in some other more flexible professions, where people give an account of the number of hours completed. This might actually lead to less wasted time and clock-watching and fewer lengthy lunch-hours and spurious site visits (some men might resent this intrusion on their freedom). Flexitime is a move in this direction which already exists in some surveying practices, but there is a need to pursue this avenue even further to enable women to reintegrate home and work (whilst possibly spreading professional work between each location). Already, many people are predicting that the large office blocks and central business districts of our cities are going to become increasingly redundant as more mundane work is computerised and more professional work is dispersed, as is possible with the assistance of home-based computer terminals, fax machines, and better telecommunication – so women should show the way. At this point in the discussion, someone always raises the question of the mythical important client who drops into the office unannounced and wants to see the woman surveyor in charge of his brief immediately. I am sure that men have encountered exactly the same problem for many years when they have been 'out' on site. Perhaps the solution is better 'management' of clients, more mobile telephones, and delegation within the office to cover such emergencies: but it is **not** specifically a woman's problem which justifies no alterations whatsoever being made to current work patterns.

A rather different situation applies to those who are employed in the public sector where everything is meant to be 'better'. There is a general feeling that the government should seek to be less hypocritical and 'put its money where its mouth is' regarding so-called 'equal opportunities'. There seems to be much 'paper provision' but still insurmountable problems for many women on a day-to-day basis. However, some local authorities are so short of professional staff that they are offering either nursery facilities themselves, or trying out various 'voucher systems' for child care (along the lines of luncheon vouchers). A particular problem would appear to be the over-bureaucratic nature of much local government which generally seems to be less flexible as to hours and attendance than in the private sector. The 'tyranny of the office' situation was such that, in some cases, those in charge preferred young women surveyors to be at their desks when there was little to do, than have them going out on professional business or having them go early to fulfil essential domestic duties (even if they volunteered to make the time up later). There simply seemed to be a fair amount of distrust and inflexibility, especially towards younger women, which actually worked against the higher goals of government organisations to be efficient, economic, and serve society. Indeed, the little power games of each department and the prevalent bureaucracy have often become ends in themselves (at the expense of the rate-payers), which has greatly shocked many keen, public-spirited women, especially those from entrepreneurial backgrounds.

Another alternative is to 'allow' women surveyors more part-time work opportunities 'so they can look after their children'; this is potentially very dangerous in the long run for women as a whole, but can be advantageous in the short term for individuals. Such work is usually lower paid, and also encourages employers to see women as a separate marginal category of worker. It may rebound on other women who do not want part-time work, and/or have no child care problems, or want to have full-time careers, or are dependent on full-time pay. The situation is complex as some women professionals see part-time work as a basic feminist right. All viewpoints and choices must be respected, but to apply solutions that are right for some women to all women can be very damaging. However, women in town planning and surveying who have pressed for this option have met with little success and have often been told, 'part-time work is only for typists, not professionals'.

Wider policy considerations

Changing the men

Much emphasis has been put on the need for women to change, to be more assertive, etc. (Lamplugh, 1988), whilst the men have often left the women to get on with it and remained unchanged. How we get the men to change their attitudes is quite another matter. Surveyors would benefit much from gaining a wider world view of the 'possibilities' by having more contact with professionals from other countries and colleagues from other professional groups which contain greater numbers of women, such as accountants, lawyers, town planners, and housing people (RTPI, 1988, for example). This might be achieved through a wider range of continuing professional development (CPD) programmes on a joint basis. The current moves to 'harmonise' and 'streamline' professional bodies in the light of both EEC and domestic government initiatives may also give some 'space' for reshaping the subculture of surveying. In particular, surveyors have arguably become much too 'narrow' and 'commercial' in their world view, although, paradoxically, the scope of their professional practice has widened and diversified. Both men and women in surveying need to become more aware of the gender and class characteristics of others (a professional duty which they cannot simply pass over to the town planners and which might reveal new markets and sources of 'man'power). Greater contact with modern management studies, both at college and CPD level, might possibly make surveyors aware of the importance of women within organisations, and of alternative approaches to interpersonal relationships in general. Since the effects of the Big Bang, competition from other professional and business organisations within the property world may actually be forcing surveyors to become less amateur and 'old-fashioned' and to re-evaluate the use of their 'man'power resources in order to compete in the modern business world. For the women's part, many simply wanted to 'be taken more seriously' and 'be valued more' as people and as surveyors.

Conclusion

There are conferences and committees run by women in the other landed professions, but it must be said that many of the people who support such ventures are either from the public sector or in academia and are women. There have been various 'networking' associations set up by women, one of the most recent being WIP (Women in Property). Such organisations are good at putting women in contact with each other, breaking down the fragmentation, helping with difficulties, and providing inspiring role models. However, there is a need for greater cross-discipline networking so that women surveyors know what women planners, geographers, and housing managers are doing, and to open their eyes to alternative ways of planning urban areas to the greater advantage of women (it is not the intention of this chapter to discuss these policy alternatives *per se*). Men surveyors love going to conferences. One woman planner said that the way to get men to attend 'women's' conferences was to charge more for them, as cost and location reflect importance in a man's world. Some believe that £300 for a day conference at the Hilton will have far more drawing power in attracting 'important' men who need 'converting' than £2.50 for the same conference in a spare lecture room in a Polytechnic, even at the cost of excluding women.

Changing the built environment

In conclusion, it is a chicken and egg situation in that more fundamental aspatial and spatial changes will be difficult to achieve without these basic facilities, but they are unlikely to be provided without a change in priorities on the part of the men. I am aware, in writing this conclusion, that it may all be wishful thinking, although some of my wishes might be achievable. A general feeling of 'hopelessness' surrounds discussions of these issues, expressed in comments such as 'you can't win', 'if they don't like you, you've had it'. But some women are achieving, and some are plodding on regardless, hoping they have not been simply 'taken for a ride' in their desire to pursue a career. There is much need (and little scope) for trying to change the whole aspatial structure of surveying in order to enable more women to go through to the top. They would provide both influence and role models for other women. Likewise, the criteria for promotion and partnership need to be re-evaluated to take into account women's 'different' attributes, skills, and experiences (as well as their 'same' attributes as men, and maybe down-grade some of the more questionable aggressive attributes that some men value), many of which may be of financial value to the business (and I don't mean their looks or 'appeal', which are relatively short-lived in any case).

Regarding spatial policy, I would like to see socially motivated town planning being reintegrated with 'developers' planning' into surveying

Effects and implications

(as was more the case in the post-war reconstruction period), provided it would have the full benefit of the influence of feminist planning insights and activities. I do, however, fully acknowledge the fact that developers expect a financial return on their investment. Social issues have been marginalised within the property development process, but a more 'feminist' development control system under the aegis of local authority planners could compensate for this lack in the private sector (provided more of them are women of course). Development control through planning law can be a very negative business, therefore there is much to be said for getting a feminist perspective built into private sector thinking. (This might be seen as too pragmatic for those who believe feminists, especially socialist feminists, should seek to avoid or abolish private property and 'the system'.) To do this one has got to convince the private sector that there is money in it and that the market demand exists. As stated, some developers and their advisers are becoming more aware of women's needs (Fitch, 1985) but they have to be reminded again and again, 'it doesn't come naturally'. There is a need for women to get into positions of influence and respect. To do this they may have to project a business image, and be twice as good as the men at 'normal' professional practice. There is already a small minority of women who have achieved seniority and are using their influence to bring about urban change and improvements in the position of women in surveying.

Changing urban theory

There is also a need to influence the nature of urban academic theory and professional literature as this influences practice. Perhaps one day gender will be a 'normal' part of urban sociological studies, fully integrated with space and class, and women's needs will also be seen as a valid aspect of 'serious' financial decision-making in the world of property development. It is important to continue research on women in the landed professions if only 'to shame them into doing something'. Likewise, women who have been given the 'woman's role' of being involved in property research in the private sector might use their humble position to make women's issues one of the integral points of reference. It is also important to continue to convince mainstream feminists that 'space matters', and to raise urban feminist issues in society. In the long run change may only be transmitted into the surveying subculture when the signal becomes strong enough from society as a whole.

Conclusion

Perhaps women have another trump card that men have dismissed in their 'other' role at home. Some women surveyors who have children have great faith in trying to bring up their sons and daughters in a non-sexist manner so that with time they might influence the surveying tribe from within the home, and perhaps ultimately, in conjunction with the efforts of other women in the other professional tribes, shape society itself and the nature of the built environment. Meanwhile, many women are 'doing their bit' in surveying education and practice. Simply 'being there' as an alternative voice is important in itself.

Appendix one

RICS membership figures, 1989

Key: LA = land agency; BS = building surveying; GP = general practice (housing is an option within this division); LS = land surveying; MS = minerals surveying; PD = planning and development; QS = quantity surveying. *Source*: RICS Records Data Base.

Table A Total membership in figures

Grade	LA	BS	GP	LS	MS	PD	QS	Total
Fellow	2,146	1,408	11,498	340	152	616	7,422	23,582
Professional Associate	2,002	3,251	14,886	562	382	740	14,553	36,376
Probationer	378	1,131	4,114	494	145	191	4,468	10,921
Student	162	1,805	3,721	240	127	332	4,285	10,672
Others	6	0	29	2	4	0	10	51
Total	4,694	7,595	34,248	1,638	810	1,879	30,738	81,602
Under 33*	819	1,350	6,216	76	91	175	4,781	13,508

Table B Female membership in figures

Grade	LA	BS	GP	LS	MS	PD	QS	Total
Fellow	5	4	72	0	0	8	15	104
Professional Associate	77	98	1,230	11	1	61	265	1,743
Probationer	72	75	873	31	5	30	284	1,370
Student	28	134	840	24	6	39	404	1,475
Others	0	0	11	0	0	0	0	11
Total	182	311	3,026	66	12	138	968	4,703
Under 33*	68	69	865	9	1	37	201	1,250

Appendices

Table C Male membership as percentages of totals

Grade	LA	BS	GP	LS	MS	PD	QS	Total
Fellow	99.8	99.7	99.3	100.0	100.0	98.7	99.8	99.5
Professional Associate	97.2	97.0	91.8	98.1	99.8	91.8	98.2	95.2
Probationer	81.0	93.4	78.8	93.8	96.6	84.3	93.7	87.5
Student	82.8	92.6	77.5	90.0	95.3	98.3	90.6	86.2
Others	100.0	–	63.0	100.0	100.0	100.0	100.0	78.5
Total	96.2	96.0	91.2	96.0	98.5	92.7	96.9	94.2
Under 33*	91.7	94.9	86.1	88.2	99.0	79.0	95.8	90.8

* Total corporate members under 33 (Fellows and Professional Associates)

201

Appendix two

Comparisons with other professions in 1989

Non-landed professions: summary

British Medical Association – 27 per cent of membership is female = 20,188 out of 76,023 total. Many of these are in the lower levels and general practice, and the 'gynaecological' areas. There are very few consultants or senior surgeons. Part-time work is catered for more, but this is a mixed blessing.

Law Society – 15 per cent of solicitors are women (there is a very slight proportionate drop in the numbers of men going into law). Only seven high court judges out of seventy-nine are female. 13 per cent of barristers are women. Many women are in 'gyny' areas such as probate, family, and welfare law, and increasingly conveyancing (Molyneux,1986). 50 per cent of law graduates and newly qualified solicitors are now women.

Institute of Chartered Accountants – Over 8 per cent of accountants are women out of a total membership of over 80,000 (comparable to the RICS). 10 per cent of all graduates from all disciplines became accountants in 1986! Such intakes must have a knock-on effect on women's chances. (See also Crompton and Sanderson, 1990: 88–108 and Chapter 5, all on women accountants.)

Other landed professions: summary

N.B. Royal Institution of Chartered Surveyors – total qualified membership (Fellows and Professional Associates) = 3 per cent women, and 97 per cent men (1,847 women of 59,958); and of the total membership 5.8 per cent are women (derived from the grand total of 94.2 per cent on Table C, Appendix 1).

Royal Town Planning Institute – 18 per cent of all members (including students) are women = 2,715 (of whom 1,099 are students) out of total membership of 14,935 (3,181 of whom are students). 15 per cent of fully qualified members are women, but the actual number in full-time employment is nearer 7 per cent. Women now make up 35 per cent of student members of the Institute. (See also Nadin and Jones, 1990.)

Royal Institute of British Architects – Women make up 8.5 per cent (nearer 4 per cent in employment) out of a membership around 28,000.

Incorporated Society of Valuers and Auctioneers – 6 per cent of members are women = 419 (266 of whom are student members) out of a total membership of 7,096. Many of the older women are also members of RICS.

National Association of Estate Agents – 14.6 per cent of membership in 1989 are women = 973 out of 6,653 total. This is not an examining body, but more of a 'trade' association.

Rating and Valuations Association – about 8 per cent and growing.

Institute of Housing – 39 per cent of the membership in 1989 are women = 3,498 (of whom 2,100 are students) out of total membership of 8,960.

Architects and Surveyors Institute – 0.7 per cent are women = 37 out of 5,230. The ASI is composed of what were the separate bodies of the Faculty of Architects and Surveyors (FAS) and the Construction Surveyors Institute (CSI) which amalgamated in 1989. Significantly, the FAS was the first surveying body to have a woman president (1987–88). (Around 1 per cent female membership is common in the smaller surveying and construction bodies which have memberships of around 4,000–6,000 each.)

Appendices

Chartered Institute of Building – less than 1 per cent = 273 out of a total of 28,382. (See also Gale, 1989b.)

Institution of Civil Engineers – around 2 per cent = 1,424, of whom 1,080 are student and graduate (probationer) members, out of a total membership of 70,118.

N.B. Everyone asks me how many black women surveyors there are. Overall I estimate there are about 0.7 per cent in all the landed professions. I have had great difficulty getting any figures on this, 'it's a very sensitive issue so we don't keep records that way'.

Appendix three

A summary of the range of courses within surveying

In view of course development and differences of designation, this list is not completely accurate, but aims to show the general distribution of full-time degree level courses (and equivalents) between the different surveying specialisms. Indeed, diversity, dispersal, modification, and upgrading are characteristics of the structure of surveying education.

Undergraduate courses

Land surveying – three minerals, two hydrographic, four land agency (rural), and seven land surveying courses (not all RICS exempting), equally spread between universities, polytechnics, and other colleges, including a leading land surveying course in a London polytechnic. Land agency might better be included with estate management for, as one prospectus proudly put it, the college had been, 'teaching the sons, and more recently the daughters, of farmers from all over the world since 1945' and 'much British rural land is controlled by former students'.

Building surveying – there are around twenty-two 'building' degrees and civil engineering and construction science courses (but only a few, so far, give substantial RICS exemptions); eight other colleges, offering between them four part-time, two full-time, two sandwich course degrees, and modularised multi-mode options (and growing), and many other external routes.

Quantity surveying – four university courses (two full-time, two sandwich); and seven technical colleges and fourteen polytechnics offering six full-time, thirteen sandwich, and eight part-time courses at degree level (and growing); and many other qualifying RICS Part I, II, and Finals courses.

Appendices

General practice (estate management) – six universities (five full-time, one sandwich), thirteen polytechnics, and several technical colleges offering between them twelve full-time, seven part-time, and five sandwich courses (and growing); also six specialised planning and development option courses, but this is also an integral part of most GP courses too. There are several other RTPI planning courses which give RICS partial exemption.

Housing – five established undergraduate courses and about ten postgraduate ones. Housing education is currently a vastly expanding area, with around 100 colleges teaching courses ranging from BTech, and 'professional qualification' day release routes through to higher degree level courses. There is also a plethora of short specialist certificate courses.

Graduate courses

There are few *postgraduate* degrees, except in high-tech land surveying, four in general practice, plus, significantly, two MBA type courses with a property option (and others planned). There are a few, and growing, number of postgraduate housing degrees. There are also numerous ongoing continuing professional development courses, many with a practical commercial emphasis.

Other routes

There many other external and part-time courses both in technical colleges or 'piggybacked' onto polytechnic surveying courses, as follows:

Building surveying – twenty-five colleges including twelve polytechnics.

Quantity surveying – forty courses, half of which are in technical colleges.

General practice – twenty-nine courses, half of which are in technical colleges.

Housing – an extensive range of technical colleges and polytechnics provide Institute of Housing part-time day release courses. Several polytechnics are in the process of offering part-time degrees.

Appendices

The CEM (College of Estate Management, Reading), and the 'private' Ellis School of Surveying and Building continue to provide a wide range of correspondence courses.

The RICS is currently phasing out the external examination route, and introducing more college-centred, modularised, part-time degrees and diplomas, and distance learning instead (mainly through CEM).

Source: RICS, July 1987a, plus many additional investigations and college prospectuses.

Bibliography

The first name of the author is included to indicate gender.

Acker, Joan (1973) 'Women and social stratification. A case of intellectual sexism', *American Journal of Sociology*, 78, Part 4: 936–45.

────── (1988) 'Class, gender, and the relations of distribution', *Signs*, 13, 3: 473–97.

Acker, Sandra (1983a) 'No woman's land: British sociology of education 1960–1979', in Ben Cosin and Margaret Hales (eds) (1983) *Education, Policy and Society*, London: Routledge and Kegan Paul, with the Open University, Milton Keynes.

────── (1983b) 'Women, the other academics', *Women's Studies International Forum*, 6, 2: 191–201.

────── (1984) 'Women in higher education: what is the problem?' in Sandra Acker and David Warren Piper (eds) *Is Higher Education Fair to Women?* Slough: National Foundation for Educational Research–Nelson.

Alford, Helen (1954) 'The maintenance and management of high density housing estates', *Journal of the Royal Institution of Chartered Surveyors*, XXXIV, Part VI: 457–72, December.

Allen, Jim (1977) 'To the lassies', JO News and Views, *Chartered Surveyor*, 109: 313, April.

Ambrose, Peter (1986) *Whatever Happened to Planning?* London: Methuen.

Ambrose, Peter and Colenutt, Barry (1979) *The Property Machine*, Harmondsworth: Penguin.

Amoo-Gottfried, Hilda (1988) 'Racism within the legal profession', *Law Society's Gazette*, 85, 1: 10–20.

Ardener, Shirley (ed.) (1978) *Defining Females: the Nature of Women in Society*, London: Croom Helm.

────── (ed.) (1981) *Women and Space, Ground Rules and Social Maps*, London: Croom Helm.

Ardrey, Robert (1967) *The Territorial Imperative*, London: Collins.

Ashworth, William (1968) *The Genesis of Modern British Town Planning*, London: Routledge and Kegan Paul.

Atkins, Susan and Hoggett, Brenda (1984) *Women and the Law*, Oxford: Blackwell.

Atkinson, Paul (1985) *Language, Structure and Reproduction*, London: Methuen.

Bibliography

Attfield, Judy (1989) 'Inside Pram Town, a case study of Harlow house interiors, 1951–61', in Judy Attfield and Pat Kirkham (eds) *A View from the Interior: Feminism, Women, and Design*, London: Women's Press.
Avis, Martin and Gibson, Virginia (1987) *The Management of General Practice Surveying Firms*, Reading: University of Reading, Faculty of Urban and Regional Studies.
Bailey, Joe (1975) *Social Theory for Planning*, London: Routledge and Kegan Paul.
Banks, Olive (1981) *Faces of Feminism: A Study of Feminism as a Social Movement*, Oxford: Martin Robertson.
Barclay, Irene (1976) *People Need Roots*, London: National Council of Social Service (Bedford Square Press).
—— (1980) 'Recollections of a Pioneer', *Chartered Surveyor*, 113, 4: 246–8, November.
Barty-King, Hugh (1975) *Scratch a Surveyor*, London: Heinemann.
Bassett, Keith and Short, John (1980) *Housing and Residential Structure, Alternative Approaches*, London: Routledge and Kegan Paul.
Battersby, E. (1970) 'Presidential Address', *Chartered Surveyor*, 103, 2: 57–61, August.
Becker, Howard, Geer, Blanche, Hughes, Everett and Strauss, Anselm (1961) *Boys in White*, Chicago: University of Chicago Press.
Benn, Caroline (1979) 'Elites versus equals', in David Rubinstein, *Education and Equality*, Harmondsworth: Penguin.
Berger, Peter (1975) *Invitation to Sociology: A Humanistic Perspective*, Harmondsworth: Penguin.
Berger, Peter and Luckman, Thomas (1972) *The Social Construction of Reality*, Harmondsworth: Penguin.
Bernard, Jessie (1981) *The Female World*, New York: Free Press.
Bernstein, Basil (1975) *Class, Codes and Control*, London: Routledge and Kegan Paul.
Best, Robin and Anderson, Margaret (1984) 'Land-use structure and change in Britain, 1971 to 1981', *The Planner*, 70, 11: 21–4, November.
Blaug, Mark (1972) *An Introduction to the Economics of Education*, Harmondsworth: Penguin.
Blowers, Andrew and Pepper, David (eds) (1987) *Nuclear Power in Crisis*, London: Croom Helm.
Blumer, Herbert (1965) 'Sociological implications of the thought of George Herbert Mead', in Ben Cosin, I. Dale, G. Esland, D. Mackinnon and D. Swift (1977) *School and Society. A Sociological Reader*, London: Routledge and Kegan Paul.
Bottomore, T. (1973) *Elites and Society*, Harmondsworth: Penguin.
Bourdieu, Pierre (1973) 'Cultural reproduction and social reproduction', in R. Brown *Knowledge, Education and Cultural Change*, London: Tavistock.
Bowlby, Sophie (1988) 'From corner shop to hypermarket: women and food retailing', in Jo Little, Linda Peake and Pat Richardson (eds) *Women and Cities, Gender and the Urban Environment*, London: Macmillan.
—— (1989) 'Gender issues and retail geography', in Sarah Whatmore and Jo Little (eds) *Geography and Gender*, London: Association for Curriculum Development in Geography.

Bibliography

Bowles, Samuel and Gintis, Herbert (1976) *Schooling in Capitalist America*, London: Routledge and Kegan Paul.
Boyd, Nancy (1982) *Josephine Butler, Octavia Hill, Florence Nightingale: Three Victorian Women who Changed the World*, London: Macmillan.
Brett-Jones Report (1978) *Review of Educational Policy*, London: RICS.
Brion, Marion and Tinker, Anthea (1980) *Women in Housing. Access and Influence*, London: Housing Centre Trust.
Broady, Maurice (1968) *Planning for People*, London: National Council of Social Service (Bedford Square Press).
Brothers, Joyce (1981) *What Every Woman Should Know About Men*, London: Granada.
Built Environment (1984) Special Issue on 'Women and the Built Environment', 10, 1.
Bulmer, Martin (1977) *Sociological Research Methods*, London: Macmillan.
────── (1984) *The Chicago School of Sociology*, London: University of Chicago Press.
Bunch, Charlotte and Pollock, Sandra (eds) (1983) *Learning our Way*, New York: Crossing Press.
Burke, Gerald (1980) *Town Planning and the Surveyor*, London: Estates Gazette.
Burke, Linda (1985) 'Changing minds: towards gender equality in science and society', paper given at British Association for the Advancement of Science, Annual Meeting, Strathclyde, Scotland, London: BAAS.
Callaway, Helen (1987) *Gender, Culture and Empire*, Urbana: University of Illinois Press.
Campbell, Beatrix (1985) *Wigan Pier Revisited: Poverty and Politics in the Eighties*, London: Virago.
────── (1987) *The Iron Ladies: Why Women Vote Tory*, London: Virago.
Carter, Ruth and Kirkup, Gill (1989) *Women in Engineering*, London: Macmillan.
Cass, Bettina (ed.) (1983) *Why so few? Women Academics in Australian Universities*, London: Sydney University Press.
Castle, Barbara (1967) 'The tramcar – urban transport of tomorrow?' *Chartered Surveyor*, 99, 11: 573–77, April.
Chadwick, Edwin (1842) *Report on the Sanitary Condition of the Labouring Population of Great Britain*, London.
Chapman, Anne (1981) 'Reviews: New Space for Women', *Women and Environments*, 4, 3: 11, Ontario: Centre for Urban and Community Studies.
Chapman, Dennis (1948) 'Social aspects of town planning', *Journal of the Royal Institution of Chartered Surveyors*, XXVIII, Part IV: 215–22, October.
Cherry, Gordon (ed.) (1981) *Pioneers in British Town Planning*, London: Architectural Press.
CIS (Counter Information Service) (1983) *The Wealthy*, London: Russell Press.
Clapham, Anthony (1949) *A Short History of the Surveyors Profession*, London: RICS.
Clark, Burton (1983) 'Belief', in *The Higher Education System: Academic Organisation in Cross-national Perspective*, Berkeley: University of

California Press.
Cline, Sally and Spender, Dale (1987) *Reflecting Men at Twice their Natural Size*, London: Andre Deutsch.
Cluttons (1987) *Cluttons, Chartered Surveyors*, London: Cluttons.
Cockburn, Cynthia (1977) *Brothers – Male Dominance and Technological Change*, London: Pluto.
────── (1985a) *The Local State: Management of People and Cities*, London: Pluto.
────── (1985b) *Machinery of Dominance*, London: Pluto.
Cohen, Philip (1976) 'Subcultural conflict and working-class community' in Martyn Hammersley and Peter Woods (eds) *The Process of Schooling*, London: Routledge and Kegan Paul, with the Open University, Milton Keynes.
Coleman, Alice (1985) *Utopia on Trial*, London: Martin Shipman.
Collins, Randall (1979) *The Credential Society*, New York: Academic Press.
Connell, Robert (1987) *Gender and Power*, Cambridge: Polity Press.
Connell, Robert, Ashenden, D., Kessler, S. and Dowsett, G. (1982) *Making the Difference: Schools, Families and Social Division*, London: Allen and Unwin.
Cooke, Philip (1987) 'Clinical inference and geographical theory', *Antipode: a Radical Journal of Geography*, 19, 1: 69–78, April.
Cooper, Cary and Davidson, Marilyn (1982) *Women in Management*, London: Fontana.
Coote, Anna (1979) *Equal at Work: Women in Men's Jobs*, London: Collins.
Corbusier, Le (1971) *The City of Tomorrow*, London: Architectural Press, first published in 1929.
Cornes, Deirdre and Lamplugh, Diana (1987) 'Suzy Lamplugh: have the lessons been learnt?' *Chartered Surveyor Weekly*, 18, 7: 5, 19.2.87.
Cox, Alan (1985) 'Cutbacks in education', *Chartered Surveyor Weekly*, 11, 4: 275, 25.4.85.
Crawford, David (1983) *Chartered Quantity Surveyor*, London: Kogan Page.
Cripps, Yvonne (1986) 'The professions: a critical review', *Law Society's Gazette*, 83, 28: 2297–301.
Crompton, Rosemary and Jones, Gareth (1984) *White-collar Proletariat, Deskilling and Gender in Clerical Work*, London: Macmillan.
Crompton, Rosemary and Mann, Michael (1986) *Gender and Stratification*, Cambridge: Polity Press.
Crompton, Rosemary and Sanderson, Kay (1990) *Gendered Jobs and Social Change*, London: Unwin Hyman.
Cross, Nigel, Elliott, David and Roy, Robin (eds) (1974) *Man Made Futures: Readings in Society, Technology and Design*, London: Hutchinson, with the Open University, Milton Keynes.
Culley, Margo, and Portuges, Catherine (1985) *Gendered Subjects, the Dynamics of Feminist Teaching*, London: Routledge and Kegan Paul.
Davidoff, Leonore and Hall, Catherine (1987) *Family Fortunes: Men and Women of the English Middle Classes, 1780–1850*, London: Hutchinson.
Davies, Brian (1976) *Social Control and Education*, London: Methuen.
Davies, Gerald (1972) *The Education of Urban Estate Surveyors*, Bristol: Bristol Polytechnic, unpublished dissertation.

Bibliography

Deem, Rosemary (ed.) (1980) *Schooling for Women's Work*, London: Routledge and Kegan Paul.
────── (ed) (1984) *Co-Education Reconsidered*, Milton Keynes: Open University.
────── (1987) 'Unleisured lives: sport in the context of women's leisure', *Women's Studies International Forum*, 10, 4: 423–32.
Delamont, Sara (1976) 'The girls most likely to: cultural reproduction and Scottish elites', *Scottish Journal of Sociology*, 1, 1: 29–43.
────── (1984) 'Dollies, debs, swots and weeds', in Geoffrey Walford (ed.) *British Public Schools: Policy and Practice*, Lewes: Falmer Press.
────── (1985) 'Fighting familiarity', *Strategies of Qualitative Research in Education*, Warwick: ESRC Summer School.
Delgado, Alan (1979) *The Enormous File: A Social History of the Office*, London: John Murray.
Delphy, Christine (translated Diana Leonard) (1984) *Close to Home: A Materialistic Analysis of Women's Oppression*, London: Hutchinson.
DES (Department of Education and Science) (1966) *A Plan for Polytechnics and other Colleges*, White Paper, Cmnd 3006, London: HMSO.
Dewar, Alison (1987) 'Social Construction of Gender in Physical Education', *Women's Studies International Forum*, 10, 4: 453–65.
Diamond, Derek (1986) 'Who gets what, where, why?' *Town and Country Planning*, 55, 3: 98.
Dickson, Anne (1982) *A Woman in Your Own Right: Assertiveness and You*, London: Quartet.
Dickson, John (1985) 'Some correlates of discretion for Chartered Surveyors', *Journal of Management Studies*, 22, 2: 213–24.
Donovan, Josephine (1985) *Feminist Theory: The Intellectual Traditions of American Feminism*, New York: Frederick Ungar.
Dore, Ronald (1976) *The Diploma Disease*, London: Allen and Unwin.
Dorfman, Marc (1986) 'Royal architecture: how it took off', *Town and Country Planning*, 55, 2: 50–3, February.
Douglas, J. (1967) *The Home and the School*, London: Panther.
Dresser, Madge (1978) Review essay of Davidoff, Leonore, L'Esperance, Jean, and Newby, Howard (1976) 'Landscape with figures: home and community in English society', *International Journal of Urban and Regional Research*, 2, 3, Special Issue on 'Women and the City'.
DTI (Department of Trade and Industry) (1988) *Rights of establishment: proposal for a council directive on a general system for the recognition of higher degree diplomas*, Information Note No. 5, London: HMSO.
Du Bois, Barbara (1983) 'Passionate scholarship: notes on values, knowing and method in feminist social science', in Gloria Bowles and Renate Duelli Klein (1983) *Theories of Women's Studies*, London: Routledge and Kegan Paul.
Dunleavy, Patrick (1980) *Urban Political Analysis*, London: Macmillan.
Durkheim, Emile (1970) *Suicide: A Study in Sociology*, London: Routledge and Kegan Paul (first published 1897).
Dworkin, Andrea (1983) *Right Wing Women*, London: Women's Press.
Edgson, P. (1969) 'Developers and tenants', *Chartered Surveyor*, 101, 8:

378–82, February.
Ekistics (1985) *Woman and Space in Human Settlements*, 52, 310, January.
Ellis, Jennifer (1973) 'Plumbing the depths', *Chartered Surveyor*, 106: 142, December.
Elston, Mary Anne (1980) 'Medicine', in Rosalie Silverstone and Audrey Ward (eds) (1980) *Careers of Professional Women*, London: Croom Helm.
Estler, Suzanne, Prussin, Labelle, Ryckman, David and Sasanoff, Robert (1985) *Gender Related Cultures and Architectural Values*, Washington: University of Washington Press.
Etzioni, Amitai (1969) *The Semi-Professions and their Organisation: Teachers, Nurses and Social Workers*, New York: Free Press.
Eve, Tristram (1948) 'Town and Country Planning Act', correspondence between Tristram Eve and Lewis Silkin, *Journal of the Royal Institution of Chartered Surveyors*, XVIII, Part IV: 204–6, October.
Eve Report (1967) *Report No. 2 of the Educational Policy Committee*, London: RICS.
Every-Woman's Encyclopaedia (1911) 'University College, Reading', Part 23: 2838–41, London: The Amalgamated Press.
Fenton-Jones, M. (1962) 'The planning of shopping centres', *Chartered Surveyor*, 95, 2: 75–80, August.
FIG (Fédération Internationale des Géomètres – International Federation of Land Surveyors) (1983) XVIII Congress, Conference Papers 202.1 – 202.7, Sophia, Bulgaria.
Finch, Janet (1983) *Married to the Job: Wives' Incorporation in Men's Work*, London: Allen and Unwin.
Firestone, Shulamith (1979) *The Dialectic of Sex*, London: Women's Press.
Fitch (1985) *Shopping Centre Report No. 8, 'Women'*, London: Fitch and Company Shopping Consortium.
Fitzherbert, John (1523) *Book of Husbandry and Book of Surveying*, London.
Foley, Donald (1964) 'An approach to urban metropolitan structure', in Melvin Webber, John Dyckman, Donald Foley, Albert Guttenberg, William Wheaton and Catherine Bauer Wurster (1964) *Explorations into Urban Structure*, Philadelphia: University of Pennsylvania Press.
Foord, Jo and Gregson, Nicky (1986) 'Patriarchy: towards a reconceptualisation', *Antipode: a Radical Journal of Geography*, 18, 2: 186–211, September.
Ford, S. (1920) *Notes on Property Law and Investment*, London: Eveleigh Nash.
Franklin, Philip (1976) 'Presidential address', *Chartered Surveyor*, 109: 135–140, December.
French, H. W. (1969) 'The pattern of further education', *Chartered Surveyor*, 101, 12: 582–94, June.
French, Marilyn (1985) *Beyond Power: Women, Men and Morals*, London: Jonathan Cape.
Freund, Otto Kahn (1978) *Selected Writings*, London: Stephens.
Gale, Andrew (1989a) 'Women in the British construction professions', Gender and Science and Technology 5th International Conference, Haifa, Jerusalem: GASAT.

Bibliography

―――― (1989b) 'Attracting women to construction', *Chartered Builder*, September/October, London: Chartered Institute of Building.
Gallese, Liz Roman (1987) *Women Like Us*, London: Grafton Books.
Gans, Herbert (1967) *The Levittowners*, London: Allen Lane.
Gardiner, A. (1923) *The Life of George Cadbury*, London: Cassell.
Gardner, Godfrey (1976) *Social Surveys for Social Planners*, Milton Keynes: Open University.
Gerstein, Martin, Lichtman, Marilyn and Barokas, Judy (1988) 'Occupational plans of adolescent women compared to men: a cross-sectional examination', *Career Development Quarterly*, 36: 222–30.
Gerth, Hans and Mills, C. Wright (1954) *Character and Social Structure: Psychology of Social Institutions*, London: Routledge and Kegan Paul.
Gibbs, Lesley (1987) 'Who designs the designers?' *WEB: Newsletter of Women in the Built Environment*, 6: 4, July.
Giddens, Anthony (1984) *New Rules of Sociological Method*, London: Hutchinson.
Gilligan, Carol (1982) *In a Different Voice: Psychological Theory and Women's Development*, Massachusetts: Harvard University Press.
Gilman, Charlotte Perkins (1979) *Herland*, London: Women's Press, first published 1915.
―――― (1921) 'Making towns fit to live in', *Century Magazine*, CII: 361–6, July.
Goffman, Erving (1969) *Presentation of Self in Everyday Life*, Harmondsworth: Penguin.
Goldsmith, Michael (1980) *Politics, Planning and the City*, London: Hutchinson.
Greater London Council (Women's Committee) (1984) *Working out the Future, Women and Jobs and the GLC*, London: GLC.
―――― (1985) *Women on the Move, GLC Survey on Women and Transport*, London: GLC.
―――― (1986) *Changing Places*, Report, London: GLC.
Greed, Clara (1984a) 'Women in surveying', *WEB Quarterly: Newsletter of Women in the Built Environment*, 1, 1: 8-11, London: WEB.
―――― (1984b) 'Whatever happened to patriarchal planning?' in *Planning*, 8.6.84, Gloucester: Ambit Publications.
―――― (1987a) 'Women and the built environment', *Working Papers in Urban Education*, London: Centre for Educational Studies, King's College, London University.
―――― (1987b) 'Can the micro be linked to the macro through ethnography?' paper given at St Hilda's Conference on Ethnography and Inequality, Oxford.
―――― (1987c) 'Forewarned is forearmed', paper given at Conference on Women and their Built Environment, Conference Report, London: Faculty of the Built Environment, South Bank Polytechnic.
―――― (1987d) 'Drains feminism', letter in *Town and Country Planning*, Special Edition on 'A Place for Women in Planning?' 56, 10: 279.
―――― (1988) 'Is more better?: with reference to the position of women chartered surveyors in Britain', *Women's Studies International Forum*, 11, 3: 187–97.

——— (1989) '"She's a good chap": women in construction', *Architects and Surveyors Institute Year Book and List of Members*, London: Highwood Publications.

——— (1990a) *The Position of Women in the Surveying Subculture: as observed in education and reflected in practice, with particular consideration of the implications for the nature of land use and development*, PhD Thesis (in preparation), Milton Keynes: Open University.

——— (1990b) 'The professional and the personal: a study of women in surveying', in Liz Stanley (ed.) *Feminist Praxis: Research, Theory and Epistemology*, London: Routledge.

Griffin, Susan (1984) *Women and Nature, The Roaring Inside Her*, London: Women's Press.

Griffiths, Dorothy (1985) 'Exclusion from technology', in Wendy Faulkner and Eric Arnold *Smothered by Invention: Technology in Women's Lives*, London: Pluto.

Hall, Peter (1977) *Containment of Urban England*, London: Allen and Unwin.

Hammersley, Martyn and Atkinson, Paul (1983) *Ethnography, Principles in Practice*, London: Tavistock.

Hansard (1986) Parliamentary Debates, House of Lords, Official Report, Monday 16 June 1986, Vol. 476, nos. 109–110: 120, reply by Lord Elton, London: HMSO.

Hanson, Michael (1983) 'Development doctors', *Chartered Surveyor Weekly*, 3, 1: 3, 7.4.83.

Hart, C. (1948) 'Modern influences on the university aspect of professional training in surveying', *Journal of the Royal Institution of Chartered Surveyors*, XXVIII, Part IV: 183–95, October.

Hart, K.M. and Hart, M.P.M. (1973) *Usill's Practical Surveying*, London: Technical Press.

Hartman, Heidi (1981) 'The unhappy marriage of Marxism and feminism', in Lydia Sargent (ed.) *Women and Revolution*, London: Pluto Press.

Harvey, David (1975) *Social Justice and the City*, London: Arnold.

Harwood, Enid (1987) 'Message from the President', *Portico, The Journal of Architecture and Surveying*, Winter, 1987: 1, Chippenham: Faculty of Architects and Surveyors.

Hayden, Dolores (1981) *The Grand Domestic Revolution: Feminist Designs for Homes, Neighbourhoods and Cities*, Cambridge, Massachusetts: MIT Press.

——— (1984) *Redesigning the American Dream*, London: Norton.

Healey, Patsy and Hillier, Jean (1988) 'Gender Issues in Planning Education', *Women and Planning Working Party*, Draft 2, 16/3/88, London: RTPI.

Healing, Oswold (1939) *Report of the Education Committee*, London: RICS.

Heap, Desmond (1973) 'Professions and professionalisation: what next?' *Chartered Surveyor*, 106: 4–9, July.

Hearn, Jeff and Parkin, Wendy (1987) *'Sex' at 'Work': the Power and the Paradox of Organisation Sexuality*, Brighton: Wheatsheaf.

Heath, Margaret (1963) 'Forty years of women', *Chartered Surveyor*, 96, 4, October.

Heine, Susanna (1987) *Women and Early Christianity*, London: SCM Press.

Bibliography

Hennicker-Heaton (1964) *Day Release*, London: Ministry of Education.
Hennig, Margaret and Jardim, Anne (1978) *The Managerial Woman*, London: Pan.
Hertz, Leah (1986) *The Business Amazons*, London: Andre Deutsch.
Hewlett, Sylvia (1988) *A Lesser Life: The Myth of Women's Liberation*, Harmondsworth: Sphere.
Hey, Valerie (1986) *Patriarchy and Pub Culture*, London: Tavistock.
Hill, William Thomson (1956) *Octavia Hill: Pioneer of the National Trust and Housing Reformer*, London: Hutchinson.
Hilland, Peter (1969) 'Letters to the editor: "Education"', *Chartered Surveyor*, 101, 12: 619, June.
Hillary, William (1985) 'Abolishing agents, urban property and leisure', *Chartered Surveyor Weekly*, 12, 2: 130, July.
Hillel, Mira Bar (1984) 'Israeli dealers: marriage brokers or agents?' *Chartered Surveyor Weekly*, 6, 10: 680, 8.3.84.
———— (1986) 'Alice in Blunderland', *Chartered Surveyor Weekly*, 15, 1: 99, 10.4.86.
Hinchcliffe, Tanis (1988) 'Women as property owners', paper given at the Women in Planning History: Theories and Applications, seminar of the Planning History Group, 12 April 1988, York: Institute of Advanced Architectural Studies.
Hirshon, Renee (ed.) (1984) *Women and Property, Women as Property: Power, Property and Gender Relations*, London: Croom Helm.
Hite, Shere (1988) *The Hite Report, Women and Love: a Cultural Revolution in Progress*, London: Viking.
Hoddell, Ian (1985) 'HP's source', *Chartered Surveyor Weekly*, 13, 7: 185, 28.11.85.
Hoddell Pritchard (1985) *The Hoddell Pritchard Story*, Bristol: Hoddell Pritchard.
Hodson, Phillip (1984) *Men*, London: Ariel Books, with the BBC.
Hoggett, Brenda and Pearl, David (1983) *The Family, Law and Society*, London: Butterworths.
Holcombe, Lee (1983) *Wives and Property*, Oxford: Martin Robertson.
Holford, William (1949) 'Civic design: an inquiry into the scope and nature of town planning', *Chartered Surveyor*, XXVIII, VIII: 403–24, February.
Hotz, Vivienne (1977) 'Life in the Lace Ghetto', *Broadsheet*, 47 (New Zealand Feminist Magazine).
Howard, Ebenezer (1960) *Garden Cities of Tomorrow*, London: Faber and Faber, first published 1898.
Howatt, Hilary (1987) 'Women in Planning – A Programme for Positive Action', *The Planner*, 73, 8: 11–12, August.
Howe, Elizabeth (1980) 'Role choices of urban planners', *Journal of the American Planning Association*, pp. 398–409, October.
Howe, Elizabeth and Kaufman, Jerome (1981) 'The values of contemporary American planners', *Journal of the American Planning Association*, pp. 266–78, July.
Howe, Louise Kapp (1978) *Pink Collar Workers: Inside the World of Women's Work*, New York: Avon.
Hurd, Geoffrey (ed.) (1978) *Human Societies: An Introduction to Sociology*,

London: Routledge and Kegan Paul.
Ingham, Mary (1984) *Men*, London: Century.
Irving, John and Martin, Ben (1985) 'Women in Radio-Astronomy – Shooting Stars?' Paper given at Annual Meeting of the British Association for the Advancement of Science, London: BAAS.
Jagger, Alison (1983) *Feminist Politics and Human Nature*, Brighton: Harvester.
JO, RICS (Junior Organisation of the RICS) (1986) *Salary Survey*, London: RICS.
———— (1988) *Salary Survey*, London: RICS.
Jones, Carol (1985) 'Sexual tyranny; male violence in a mixed secondary school', in Gaby Weiner (ed.) *Just a Bunch of Girls*, Milton Keynes: Open University.
Joseph, Martin (1978) 'Professional values, a case study of professional students in a Polytechnic', *Research in Education*, 19: 49-65, May.
———— (1980) *Professional socialisation: a case study of estate management students*, unpublished DPhil, Oxford University.
———— (1988) *Sociology for Everyone*, Cambridge: Polity Press.
Kanter, Rosabeth Moss (1977) *Men and Women of the Corporation*, New York: Basic Books.
————(1984) *The Change Masters*, London: Allen and Unwin.
Keeble, Lewis (1956) 'Cost of residential estate development in relation to density', *Chartered Surveyor*, LXXXVIII, II: 590–94, May.
———— (1969) *Principles and Practice of Town and Country Planning*, London: Estates Gazette.
Keller, Susan (1981) *Building for Women*, Massachusetts: Lexington Books.
Kelly, Alison (ed.) (1987) *Science for Girls?* Milton Keynes: Open University.
Kirk, Gwyneth (1980) *Urban Planning in a Capitalist Society*, London: Croom Helm.
Kleinman, Sherryl (1987) 'Women in seminary: dilemmas of professional socialisation', *Sociology of Education*, 57: 210–19.
Knight, Stephen (1985) *The Brotherhood*, London: Panther.
Knox, Paul (ed.) (1988) *The Design Professions and the Built Environment*, London: Croom Helm.
Korda, Michael (1974) *Male Chauvinism, How it works*, London: Barrie and Jenkins.
La Rouche, Janice and Ryan, Regina (1985) *Strategies for Women at Work*, London: Counterpoint.
Lamplugh, Diana (1988) *Beating Aggression: A Practical Guide for Working Women*, London: Weidenfeld and Nicolson.
Lane, L.W. (1958) 'Ten years of planning', *Chartered Surveyor*, 90, 12: 652–7, June.
Lane, Michael (1975) *Design for Degrees: A History of CNAA from its inception in 1964*, London: Macmillan.
Langdon, Horace (1949) 'The future of the profession' *Journal of the Royal Institution of Chartered Surveyors*, XXIX, Part III: 196–214, September.
Langland, Elizabeth and Gove, Walter (eds) (1981) *A Feminist Perspective in the Academy: the Difference it makes*, Chicago: University of Chicago Press.

Bibliography

Law Society (1988) *Equal in the Law: Report of the Working Party on Women's Careers*, London: The Law Society.

Lawless, Paul (1981) *Britain's Inner Cities*, London: Macmillan.

Leach, Penelope (1979) *Who cares? A New Deal for Mothers and their Small Children*, Harmondsworth: Penguin.

Leevers, Kate (1986) *Women at Work In Housing*, London: HERA.

Leoff, Constance (1987) *Bluff Your Way in Feminism*, London: Ravette.

Levin, Ira (1974) *Stepford Wives*, London: Pan.

Levison, Debra and Atkins, Julia (1987) *The Key to Equality: The Women in Housing Survey*, Women in Housing Working Party, London: Institute of Housing.

Lewis, Roy and Maude, Angus (1953) *Professional People in England*, Harvard: Harvard University Press.

Leybourn, William (1653) *The Compleat Surveyor*, London.

Little, Jo, Peake, Linda and Richardson, Pat (1988) *Women and Cities, Gender and the Urban Environment*, London: Macmillan.

Loban, Glenys (1978) 'The influence of the school on sex stereotyping' in Jane Chetwynd and Oonagh Harnett (eds) *The Sex Role System: Sociological and Psychological Perspectives*, London: Routledge and Kegan Paul.

Lorber, Judith (1984) *Women Physicians*, London: Tavistock.

LPAS (London Planning Aid Service) (1986a) *Planning for Women: An Evaluation of Local Plan consultation by three London Boroughs*, Research Report No. 2, London: Town and Country Planning Association.

—— (1986b) *Planning Advice for Women's Groups*, Community Manual No. 6, London: Town and Country Planning Association.

Lury, Celia (1987) *The Difference of Women's Writing: Essays on the Use of Personal Experience*, Manchester: University of Manchester, Studies in Sexual Politics.

Maccoby, Eleanor (1972) 'Sex differences in intellectual functioning', in S. Anderson *Sex Differences and Discrimination in Education*, London: Wadsworth.

MacDonald, Madeleine (1981) 'Schooling and the reproduction of class and gender relations', in Roger Dale, Geoff Esland, Ross Ferguson and Madeleine MacDonald *Education and the State: Volume II, Politics, Patriarchy and Practice*, Lewes: Falmer Press, with the Open University, Milton Keynes.

McDowell, Linda (1983) 'Towards an understanding of the gender division of urban space', *Environment and Planning D: Society and Space*, 1: 59–72.

—— (1986) 'Beyond patriarchy: a class-based explanation of women's subordination', *Antipode: a Radical Journal of Geography*, 18, 3: 311–21, London: Blackwell.

McDowell, Linda and Peake, Linda (1989) 'Women in British geography revisited: or the same old story?' *Journal of Geography in Higher Education* (forthcoming) from Conference of Women and Geography Study Group, Institute of British Geographers, Equal Opportunities in Geography, London.

Macey, J. (1958) 'Slum clearance – guerilla tactics or total war', *Chartered Surveyor*, 90, 8: 438–41, February.

MacFarlane, Alan (1978) *The Origins of English Individualism, the Family, Property and Social Transition*, Oxford: Blackwell.
McLoughlin, John (1969) *Urban and Regional Planning: A System's View*, London: Faber.
Mahoney, Pat (1985) *Schools for the Boys?* London: Hutchinson.
Marcus, Susanna (1971) 'Planners – who are you?' *Journal of the Royal Town Planning Institute*, 57, 2.
Marks, Peter (1988) *Solicitors' Career Structure Survey*, London: Polytechnic of Central London and Law Society.
Markusen, Anne (1981) 'City spatial structure, women's household work and national urban policy', in Catherine Stimpson, Elsa Dixler, Martha Nelson and Kathryn Yatrakis (eds) *Women and the American City*, Chicago: University of Chicago Press.
Marriot, Oliver (1989) *The Property Boom*, London: Abingdon.
Marshall, Judi (1984) *Women Managers: Travellers in a Male World*, London: Wiley.
Marwick, Ann (1986) 'Letters: women and work', *Chartered Surveyor Weekly*, 16, 7: 466, 14.8.86.
Marx, Karl (Translated A. Miller) (1981) *Grundrisse*, Harmondsworth: Penguin (first published 1857).
Massey, Doreen (1984) *Spatial Divisions of Labour: Social Structures and the Geography of Production*, London: Macmillan.
Matrix (Women Architects) (1984) *Making Space, Women and the Man Made Environment*, London: Pluto.
Mead, Margaret (1949) *Male and Female*, London: Gollancz.
Menzies, Colin (1985) 'Man of Manor Place', *Chartered Surveyor Weekly*, 12, 8: 497, 22/29.8.85.
Merrett, Stephen (1979) *Owner Occupation in Britain*, London: Routledge and Kegan Paul.
Merton, Robert (1952) 'Bureaucratic structure and personality', in Robert Merton (ed.) *Reader in Bureaucracy*, New York: Free Press.
Metcalf, Andy and Humphries, Martin (eds) (1985) *The Sexuality of Men*, London: Pluto.
Midland (1979) *Midland Bank Trust Company v. Green*, Vol. I: 496, London: Chancery Law Reports.
Midwinter, Eric (1972) *Priority Education*, Harmondsworth: Penguin.
Miller, Daniel and Swanson, Guy (1958) *The Changing American Parent: a Study in the Detroit Area*, New York: Wiley.
Millerson, Geoffrey (1964) *The Qualifying Associations*, London: Routledge and Kegan Paul.
Millet, Kate (1985), *Sexual Politics*, London: Virago, first published 1969.
Mills, C. Wright (1959) *The Power Elite*, Oxford: Oxford University Press.
—— (1978) *The Sociological Imagination*, Harmondsworth: Penguin.
Mills, Peter and Oliver, John (1967) *The Survey of the Building Sites in the City of London: after the Great Fire of 1666*, London: London Topographical Society.
Ministry of Education (1966) *A Plan for Polytechnics and other Colleges*, White Paper, London: HMSO.
Mitchell, Juliet (1981) *Woman's Estate*, Harmondsworth: Penguin.

Bibliography

Molyneux, Pauline (1986) 'Association of Women Solicitors – membership survey', *Law Society's Gazette*, 16.10.86: 3082.

Morgan, David (1981) 'Men, masculinity and the process of sociological enquiry', in Helen Roberts (ed.) *Doing Feminist Research*, London: Routledge and Kegan Paul.

Morgan, Peter and Nott, Susan (1988) *Development Control: Policy into Practice*, London: Butterworths.

Morphet, Janice (1983) 'Planning and the majority – women', paper given at the Town and Country Planning School, St Andrews, Scotland, London: Royal Town Planning Institute.

Morris, Eleanor (1986) 'An overview of planning for women from 1945–1975', in Marion Chalmers (ed.) *New Communities: Did They Get It Right?* Report of a Conference of the Women and Planning Standing Committee of the Scottish Branch of the Royal Town Planning Institute, County Buildings, Linlithgow, London: RTPI.

Morris, Terence (1958) *The Criminal Area*, London: Routledge and Kegan Paul.

Mulford, Wendy (1986) 'In this process, I too am subject', in Denise Farran, Sue Scott and Liz Stanley (eds) *Writing Feminist Biography*, Manchester: University of Manchester, Studies in Sexual Politics.

Mumford, Lewis (1930s) *The City*, American Institute of Planners film by RKO, commentary written by Mumford, shown 8.8.87 on BBC2.

—— (1965) *The City in History*, Harmondsworth: Penguin.

NAB (National Advisory Body, Department of Education and Science) (1987) 'Policy considerations', *NAB Bulletin*, Autumn, London: HMSO.

Nadin, Vincent and Jones, Sally (1990) 'A profile of the profession', *The Planner*, 73, 3: 13–24, London: RTPI.

Napier, Mary (1959) *Woman's Estate*, London: Rupert Hart-Davis.

Nationwide (1986) *Lending to Women – Nationwide*, Background Bulletin, London: Nationwide Building Society.

NCVQ (National Council for Vocational Qualifications) (1987) *The Royal Institution of Chartered Surveyors, Observations on the Consultative Document from the NCVQ*, 28.10.87, London: RICS.

Nix, John, Hill, Paul and Williams, Nigel (1987) *Land and Estate Management*, Chichester: Packard.

Norton-Taylor, Richard (1982) *Whose Land is it Anyway?* Wellingborough: Turnstone.

Nott, Susan (1989) 'Women in the law', *New Law Journal*, 139, 6410: 749–52, 2.6.89; continued 139, 6411: 785–6, 9.6.89.

Oakley, Ann (1980) *Women Confined: Towards a Sociology of Childbirth*, Oxford: Martin Robertson.

Okely, Judith (1978) 'Privileged, schooled and finished: boarding school education for girls', in Shirley Ardener (ed.) *Defining Females: the Nature of Women in Society*, London: Croom Helm.

OPCS (Office of Population Censuses and Surveys) (1983) 'Usually resident population: age by marital status by sex', *Census 1981: National Report – Great Britain, Part I*, Table 6, page 15, London: HMSO.

Orchard-Lisle, Paul (1985) 'Professionals in the '80s', *Chartered Surveyor Weekly*, 13, 7: 593–5, 14.11.85.

Pahl, Ray (1965) *Urbs in Rure*, London: Weidenfeld and Nicolson.
———— (1977a) 'Managers, technical experts and the state', in M. Harloe (ed.) *Captive Cities*, London: Wiley.
———— (1977b) 'Playing the rationality game', in Colin Bell and Howard Newby (eds) *Doing Sociological Research*, London: Allen and Unwin.
———— (1984) *Divisions of Labour*, Oxford: Blackwell.
Pahl, Ray and Pahl, Jan (1971) *Managers and their Wives*, Harmondsworth: Penguin.
Parkin, Frank (1979) *Marxism and Class Theory: A Bourgeois Critique*, London: Tavistock.
Pearce, Lynn (1988) *The Architectural and Social History of Co-operative Living*, London: Macmillan.
Percy Committee (1970) *Report of the Committee on Higher Technical Education*, Ministry of Education, London: HMSO.
Perrot, Roy (1968) *The Aristocrats*, London: Weidenfeld and Nicolson.
Perry, Elisabeth Israels (1987) *Moskowitz: Feminine Politics and the Exercise of Power in the Age of Alfred E. Smith*, Oxford: Oxford University Press.
Pevsner, Nikolaus (1970) *Pioneers of Modern Design*, Harmondsworth: Penguin.
Pickup, Laurie (1984) 'Women's gender role and its influence on their travel behaviour', *Built Environment*, Special Issue on 'Women and the Built Environment', 10, 1: 61–8.
Pickvance, Christopher (ed.) (1977) *Urban Sociology*, London: Tavistock.
———— (1987) 'Book Reviews', *International Journal of Urban and Regional Research*, 11, 2: 288.
Pilkington Committee (1966) *Technical College Resources*, Ministry of Education, London: HMSO.
Pinch, Stephen (1985) *Cities and Services: the Geography of Collective Consumption*, London: Routledge and Kegan Paul.
Potter, Stephen (1986) 'Car tax concessions; perk or problem?' *Town and Country Planning*, 55, 6: 169–76.
Power, Anne (1987) *Property Before People: the Management of Twentieth-Century Council Housing*, London: Allen and Unwin.
Purvis, June (1987) 'Understanding personal accounts', in Gaby Weiner and Madeleine Arnot (eds) *Gender under Scrutiny: New Enquiries in Education*, London: Hutchinson, with the Open University, Milton Keynes.
Quoin (1967) 'Perspective: . . . and even brighter girls', *Chartered Surveyor*, 100, 5: 227, November.
Rapoport, Rhona and Rapoport, Robert (1971) *Dual Career Families*, Harmondsworth: Penguin.
Ravetz, Alison (1980) *Remaking Cities*, London: Croom Helm.
Richards, A. (ed.) (1977) *Sigmund Freud: Case Histories, Dora and Little Hans*, Harmondsworth: Penguin.
Richardson, B. (1876) *Hygenia, A City of Health*, London.
RICS (1939) *Register of Chartered Surveyors, Chartered Land Agents, and of Auctioneers and Estate Agents*, London: Thomas Skinner Publishers.
———— (1966) 'The role of the chartered surveyor in town and country planning: report of the RICS working party', *Chartered Surveyor*, 98, 10: 537–40.

Bibliography

──────── (1986a) *Warning: Not using a Chartered Surveyor can put you at risk*, London: RICS.

──────── (1986b) *Professions in Crisis: New Opportunities for Chartered Surveyors*, Annual Conference, Cardiff, London: RICS.

──────── (1987a) *Chartered Surveying: Do you measure up?* London: RICS.

──────── (1987b) *Structure and Chartered Designations Review*, London: RICS.

──────── (1989) *The Royal Institution of Chartered Surveyors Directory*, London: Macmillan.

Riley, Mary and Bailey, Christine (1983) 'Learning to get by in a man's world', *Planning*, 543, 4.11.83, Gloucester: Ambit.

Robbins Report (1983) *Higher Education*, Ministry of Education, London: HMSO.

Roberts, Patricia (1988) 'Women and planning history: theories and applications', paper given at Women in Planning History: Theories and Applications seminar of the Planning History Group, York: Institute of Advanced Architectural Studies.

Robinson, C. (1987) 'Where have all the young ones gone? An analysis of the recruitment crisis', *Law Society's Gazette*, 84, 12: 875.

Robinson, Eric (1968) *The New Polytechnics*, Harmondsworth: Penguin.

Rowbotham, Sheila (1973) *Woman's Consciousness: Man's World*, Harmondsworth: Penguin.

──────── (1974) *Women, Resistance and Revolution*, Harmondsworth: Penguin.

Rowe, Michael (1955) 'Development plans: objects and objections', *Journal of the Royal Institution of Chartered Surveyors*, XXXIV, Part IX: 759-65, March 1955.

RTPI (Royal Town Planning Institute) (1984) *Sample Survey of Members 1984, Interim Results*, London: RTPI.

──────── (1987) *Report and Recommendations of the Working Party on Women and Planning*, London: RTPI.

──────── (1988) *Managing Equality: the Role of Senior Planners*, Conference, 28 October 1988, London: RTPI.

Saffioti, Heleieth (1978) *Women in Class Society*, New York: Monthly Review Press.

Saks, Mike (1983) 'Removing the blinkers, a critique of recent contributions to the sociology of the professions', *Sociological Review*, pp. 2–21, February.

Sarup, Madan (1978) *Marxism and Education*, London: Routledge and Kegan Paul.

Saunders, A. Carr and Wilson, P. (1933) *The Professions*, Oxford: Clarendon Press.

Saunders, Peter and Williams, Peter (1988) 'The constitution of the home: towards a research agenda', *Housing Studies*, 3, 2.

Savage, Wendy (1986) *A Savage Enquiry: Who Controls Childbirth?* London: Virago.

Sayer, Andrew (1983) 'Realism and geography', in R.J. Johnston (ed.) *The Future of Geography*, London: Methuen.

SBP (South Bank Polytechnic) (1987) *Women and their Built Environment*, conference in the Faculty of the Built Environment, South Bank Polytechnic, London: SBP.

Schuster Committee (1950) *Report on the Qualifications of Planners*, Cmd. 8059, London: HMSO.

Sharp, Rachel and Green, Anthony (1975) *Education and Social Control*, London: Routledge and Kegan Paul.

Sharpe, Sue (1976) *Just like a Girl: How Girls Learn to be Women*, Harmondsworth: Penguin.

Sheffield City Polytechnic (1986) *Final Report of the Working Party on the Place of Women in the Polytechnic*, Sheffield: Sheffield Polytechnic.

Shepherd, W. (1954) 'The reconstruction of the city centre of Plymouth', *Journal of the Royal Institution of Chartered Surveyors*, XXXIII, Part XII: 934–40, June.

Siltanen, Janet and Stanworth, Michelle (1984) *Women and the Public Sphere*, London: Hutchinson.

Silverstone, Rosalie and Ward, Audrey (eds) (1980) *Careers of Professional Women*, London: Croom Helm.

Simmie, James (1974) *Citizens in Conflict, The Sociology of Town Planning*, London: Hutchinson.

——— (1981) *Power, Property and Corporatism*, London: Macmillan.

SITE (1980) *Surveying in the Eighties*, London: RICS.

Skeffington, A. (1969) *People and Planning*, London: HMSO.

Smart, Carol (1984) *The Ties that Bind: Law, Marriage and the Reproduction of Patriarchal Relations*, London: Routledge and Kegan Paul.

Smith, Mary E.H. (1967) 'Letters to the editor: . . . and even brighter girls', *Chartered Surveyor*, 100, 6: 282, December.

Smith, Mary (1989) *Guide to Housing*, London: Housing Centre Trust.

Smith, Neil and Williams, Peter (1986) *Gentrification of the City*, London: Allen and Unwin.

Smyth, Penny (1980) 'What it's like being a woman chartered surveyor', *Chartered Surveyor Weekly*, 113: 351, December.

Snow, Joyce (1977) 'Women in the Profession', *Chartered Surveyor*, 109: 211, January.

Sousby, Jane (1977) 'Women in the surveying profession', *Chartered Surveyor*, 109: 376.

Spencer, Anne and Podmore, David (1987) *In a Man's World: Essays on Women in Male-dominated Professions*, London: Tavistock.

Spender, Dale and Spender, Lynn (1983) *Gatekeeping: Denial, Dismissal and Distortion of Women*, London: Pergamon.

Stacey, Margaret and Price, Marion (1981) *Women, Power and Politics*, London: Tavistock.

Stanley, Liz (1987) 'Some notes on "hidden" work in public places: the case of Rochdale', in Liz Stanley *Essays on Women's Work and Leisure and 'Hidden' Work*, Manchester: University of Manchester, Studies in Sexual Politics.

Stanley, Liz and Wise, Sue (1983) '"Back into the personal" or: our attempt to construct "feminist research"', in Gloria Bowles and Renate Duelli Klein (eds) *Theories of Women's Studies*, London: Routledge and Kegan Paul.

Stanworth, Michelle (1984) *Gender and Schooling: A Study of Sexual*

Divisions in the Classroom, London: Hutchinson.
Stapleton, Tim and Netting, Roger (1986) 'Not all theory', *Estates Gazette*, 8.3.86, 277: 933–4.
Steel, Robert (1960) 'The evolution of the chartered surveyor in society', *Chartered Surveyor*, 93, 4: 176–83, October.
Stewart, Katie (1981) 'The marriage of capitalist and patriarchal ideologies: meanings of male bonding and male ranking in US culture', in Lydia Sargent (ed.) *Women and Revolution*, London: Pluto Press.
Stimpson, Catherine, Dixler, Elsa, Nelson, Martha and Yatrakis, Kathryn (eds) (1981) *Women and the American City*, Chicago: University of Chicago Press.
Stone, John (1983) 'Couples are out of order', *Planning*, 546: 2, 25.11.83.
Strauss, Anselm (ed.) (1968) *The American City*, London: Allen Lane.
Strutt and Parker (1985) *The First Hundred Years*, London: Strutt and Parker.
Sturge, John Player (1986) *J.P. Sturge and Sons, 225th Anniversary*, Bristol: Sturge.
Swenarton, Mark (1981) *Homes fit for Heroes*, London: Heinemann.
Swords-Isherwood, Nuala (1985) 'Women in British engineering', in Wendy Faulkner and Erik Arnold (eds) (1985) *Smothered by Invention: Technology in Women's Lives*, London: Pluto.
Taylor, Beverley (1988) 'Organising for change within local authorities: how to turn ideas into action to benefit women', paper given at Women and Planning: Where Next?, short course at Polytechnic of Central London, 16 March 1988, London: PCL.
TCP (Town and Country Planning Association) (1987) Special Edition on 'A Place for Women in Planning?' *Town and Country Planning Journal*, 56, 10: 279.
Teale, Mark (1985) 'Chameleon profession pressurised to follow suit', *Estates Times*, 9.8.85.
Thomas, Kim (1986) *Being in a minority: the experience of first year women physics students and first year male arts students*, paper given at Aston Management Centre, February.
Thompson, F. Michael, L. (1963) *English Landed Society*, London: Routledge and Kegan Paul.
────── (1968) *Chartered Surveyors, the Growth of a Profession*, London: Routledge and Kegan Paul.
Thompson, Jane (1983) *Learning Liberation: Women's Response to Men's Education*, London: Croom Helm.
Torre, Susana (ed.) (1977) *Women in American Architecture: A Historic and Contemporary Perspective*, New York: Whitney Library of Design.
Trepas, B. (1970) 'Twilight to daylight', *Chartered Surveyor*, 103, 3: 124–31.
Turner, Jessie (1986) 'Letters: "Childrearing and surveying"', *Chartered Surveyor Weekly*, 16, 3: 172, 17.7.86.
Vaizey, John (1970) *Education?: for Tomorrow*, Harmondsworth: Penguin.
Valin Pollen (1986) *Report on the Role of the Royal Institution of Chartered Surveyors*, London: Consensus Research Limited.
Vallance, Elizabeth (1979) *Women in the House*, Athlone Press, London.
Venables, P. (1955) *Technical Education: its Aims, Organisation and Future Development*, London: G. Bell and Sons.

Bibliography

Wagner, Philip (1984) 'Suburban Landscapes and Nuclear Families', *Built Environment*, 10, 1, Special Issue on 'Women and the Built Environment'.

Walby, Sylvia (1986) *Patriarchy at Work*, Cambridge: Polity Press.

Walford, Geoffrey (1986) *Life in Public Schools*, London: Methuen.

Walker, Lynn (1989) 'Women and architecture', in Judy Attfield and Pat Kirkham (eds) *A View from the Interior: Feminism, Women, and Design*, London: The Women's Press.

Walsh, Linda and Gibson, Mike (1985) 'The average planner and CPD', *The Planner*, 71, 3.

Ward, Colin (1987) 'The lady tracers', *Town and Country Planning*, 56, 10: 255–6.

Ward, Dorcas (1963) 'The work of a housing manager', *Chartered Surveyor*, 95, 7: 392, January.

Ware, Vron (1987) 'Problems with design improvements at home', *Town and Country Planning*, 56, 10.

Wareing, Nathaniel (1986) *Professional Education: An Undergraduate's Viewpoint*, Bristol: Bristol Polytechnic, unpublished thesis.

Watson Committee (1950) *Report on the Educational Policy of the RICS*, London: RICS.

Watson, John (1977) *Savills, A Family and a Firm*, London: Hutchinson Benham.

Weber, Max (1947) *The Methodology of the Social Sciences*, New York: Free Press.

—— (1964) *The Theory of Social and Economic Organisation (Wirtschaft und Gesellschaft)*, New York: Free Press.

Weiner, Gaby (ed.) (1985) *Just a Bunch of Girls*, Milton Keynes: Open University.

Weisman, Leslie and Birkby, Noel (1983) 'The women's school of planning and architecture', in Charlotte Bunch and Sandra Pollock *Learning our Way*, New York: Crossing Press.

Wekerle, Gerda, Peterson, Rebecca and Morley, David (eds) (1980) *New Space for Women*, Boulder: Westview Press.

Wells Committee (1960) *Report on the Educational Policy of the RICS*, London: RICS.

Westcott, Marcia (1981) 'Women's studies as a strategy for change: between criticism and vision', in Gloria Bowles and Renate Duelli Klein (eds) *Theories of Women's Studies*, London: Routledge and Kegan Paul.

Westergaard, John and Resler, Henrietta (1978) *Class in Capitalist Society*, Harmondsworth: Penguin.

WGPW (Women's Group on Public Welfare, Hygiene Committee) (1943) *Our Towns: A Close Up*, Oxford: Oxford University Press, with National Council of Social Service, London.

WGSG (Women and Geography Study Group, Institute of British Geographers) (1984) *Geography and Gender*, London: Hutchinson.

Whelan, Christine (1984) 'Partners in prime', *Chartered Surveyor Weekly*, 8, 6: 362, 9.8.84.

Whitburn, Julia (1976) *People in Polytechnics*, London: Society for Research into Higher Education.

White, Julie (1987) *Image and Self-Projection for Today's Professional*

Bibliography

Woman, Slough: CareerTrack Inc.

Whitelegg, Elizabeth, Arnot, Madeleine, Bartels, Else, Beechey, Veronica, Birke, Lynda, Himmelweit, Susan, Leonard, Diana, Ruehl, Sonja and Speakman, Mary Anne (eds) (1982) *The Changing Experience of Women*, Oxford: Basil Blackwell with the Open University.

Whittington, Eric (1987) *Survey of Polytechnic Degrees*, London: Careers Consultants.

Whyld, June (1983) *Sexism in the Secondary Curriculum*, London: Harper and Row.

Whyte, Judith (1986) *Girls into Science and Technology*, London: Routledge and Kegan Paul.

Whyte, Judith, Deem, Rosemary, Kant, Lesley and Cruickshank, Maureen (eds) (1985) *Girl Friendly Schooling*, London: Methuen.

Whyte, William (1963) *The Organisation Man*, Harmondsworth: Penguin.

Wigfall, Valerie (1980) 'Architecture', in Rosalie Silverstone and Audrey Ward (eds) *Careers of Professional Women*, London: Croom Helm.

Williams Grey, Marcelle (1977) *The New Executive Woman*, New York: Mentor.

Willis, Arthur (1946) *To be a Surveyor*, London: Methuen.

Willis, Paul (1977) *Learning to Labour: How Working Class Kids get Working Class Jobs*, Aldershot: Gower.

Wilson, Elizabeth (1980) *Only Half Way to Paradise*, London: Tavistock.

Woodhall, Maureen (1972) *Economic Aspects of Education*, Slough: National Foundation for Educational Research.

Woods, Peter (1987) *Inside Schools: Ethnography in Educational Research*, London: Routledge and Kegan Paul.

Woodward, Diana (1973) *The Marriage and Career of Women Undergraduates*, unpublished PhD thesis, University of Cambridge.

Wright, Olin Erik (1985) *Classes*, London: Verso.

Young, Michael (1958) *The Rise of Meritocracy*, Harmondsworth: Penguin.

Name index

Acker, J. 23
Acker, S. 5, 13, 29, 34
Alford, H. 76
Allen, J. 79
Ambrose, P. 7, 24, 175; Ambrose, P. and Colenutt, B. 8
Amoo-Gottfried, H. 150
Antipode 27
Ardener, S. 23, 44, 106
Ardrey, S. 44
Ashworth, W. 54, 60
Atkins, S. and Hoggett, B. 45
Atkinson, P. 30, 33
Attfield, J. 67
Avis, M. and Gibson, G. 37, 124

Bailey, J. 27, 69
Banks, O. 54
Barclay, I. 61
Barty-King, H. 46
Bassett, K. and Short, J. 7, 52, 175
Battersby, E. 59, 74, 188
Becker, H. 8, 25, 35, 90
Benn, C. 32
Berger, P. 26; Berger, P. and Luckman, T. 11, 24, 56
Bernard, J. 49
Bernstein, B. 26, 30, 33, 58, 176
Best, R. and Anderson, M. 44
Blaug, M. 31
Blowers, A. 14
Blumer, H. 25
Bottomore, T. 31
Bourdieu, P. 33
Bowles, S. and Gintis, H. 170

Bowlby, S. 170
Boyd, N. 53–4
Brett-Jones 71, 74, 190
Brion, M. and Tinker, A. 54
Broady, M. 69
Brothers, J. 24
Built Environment 27, 80
Bulmer, M. 13
Bunch, C. and Pollock, S. 34
Burke, G. 66
Burke, L. 37

Callaway, H. 32
Campbell, B. 49, 53
CareerTrack 38
Carter, R. and Kirkup, G. 4
Cass, B. 34
Castle, B. 77
Chapman, A. 164
Chapman, D. 66
Cherry, G. 51, 55
CIS 44
Cline, S. and Spender, D. 115
Cluttons 46
Cockburn, C. 27, 47, 156
Cohen, P. 29
Coleman, A. 83, 162
Collins, R. 30, 91
Community Action 162
Connell, R. 23, 32
Cooke, P. 13
Cooper, C. and Davidson, M. 38
Coote, A. 38
Cornes, D. and Lamplugh, D. 84
Cox, A. 85

Name index

Crompton, R. and Jones, G. 25, 173; Crompton, R. and Mann, M. 25, 147; Crompton, R. and Sanderson, K. 18, 202
Cross, N. 27
Culley, M. and Portugese, C. 34

Davidoff, L. and Hall, C. 53
Davies, B. 26, 33
Davies, G. 34, 73, 99
Deem, R. 29, 33, 169
Delamont, S. 26, 28, 30, 32, 33, 46
Delgado, A. 25
Delphy, C. 22
DES 34
Dewar, A. 37
Diamond, D. 4
Dickson, A. 39
Dickson, J. 37
Donovan, J. 20, 22
Dore, R. 31, 91
Dorfman, M. 17
Douglas, J. 32
Dresser, M. 27
DTI 8
Du Bois, B. 13
Dunleavy, P. 3, 24, 27, 69
Durkheim, E. 53

Edgson, P. 79
Ellis, J. 79
Elston, M. 30, 37, 81
Estler, S. 37
Etzioni, A. 32
Eve, T. 9; Eve Report 72
Every-Woman's Encyclopaedia 59

Fenton-Jones, M. 76
FIG 7, 78, 188
Finch, J. 33
Firestone, S. 23
Fitch 173, 198
Fitzherbert, J. 43
Foley, D. 5
Foord, J. and Gregson, N. 27
Ford, S. 56
Franklin, P. 79
French, H. 72

French, M. 27, 188
Freund, O. 28

Gale, A. 6, 204
Gallese, L. 12, 38, 129
Gans, H. 136
Gardiner, A. 51
Gardner, G. 14
Gerstein, M. 37
Gerth, H. and Mills, C. 31, 35, 49
Gibbs, L. 10
Giddens, A. 28, 186
Gilligan, C. 177
Gilman, C. 54
Ginsberg, L. 163
GLC 27, 168, 173
Goffman, E. 28, 44, 78
Goldsmith, M. 69
Greed, C. 3, 9, 10, 12, 27, 28, 35, 36, 54, 67, 80, 86, 91, 167
Griffin, S. 27
Griffiths, D. 37, 50

Hall, P. 67
Hammersley, M. and Atkinson, P. 13–15, 29, 34
Hansard 135, 142
Hanson, M. 85
Hart, C. 59
Hart, K. and Hart, M. 15
Hartman, H. 23
Harvey, D. 69
Harwood, E. 44
Hayden, D. 3, 27, 54
Healey, P. and Hillier, J. 94
Healing Report 64, 70
Heap, D. 59
Hearn, J. and Parkin, W. 16, 25, 38, 147
Heath, M. 76
Heine, S. 46, 83, 111
Hennicker-Heaton Report 72
Hennig, M. and Jardim, A. 38
Hertz, L. 9, 32
Hewlett, S. 23
Hey, V. 109
Hill, W. 55, 63
Hilland, P. 78

Name index

Hillel, M. 45, 83
Hinchcliffe, T. 54, 178
Hirschon, R. 28
Hite, S. 23, 24
Hoddell, I. 46; Hoddell Pritchard 47
Hodson, P. 24
Hoggett, B. and Pearl, D. 52
Holcombe, L. 27
Holford, W. 66
Hotz, V. 52
Howard, E. 54
Howatt, H. 4
Howe, E. 25, 37; Howe, E. and Kaufman, J. 37
Hurd, G. 37

Ingham, M. 24
Irving, J. and Martin, B. 37

Jagger, A. 20
JO 7, 9, 79–82, 123–4, 126–7, 130, 132, 136, 157, 190
Jones, C. 33
Joseph, M. 7, 17, 25, 31, 32, 34, 44, 94, 96–8, 104, 120, 178, 194

Kanter, R. 4, 37, 38
Keeble, L. 66
Keller, S. 27
Kelly, A. 100
Kirk, G. 7
Kleinman, S. 37
Knight, S. 47
Knox, P. 37
Korda, M. 24

La Rouche, J. and Ryan, R. 4, 23
Lamplugh, D. 84, 154, 196
Lane, M. 73
Lane, W. 66
Langdon, H. 70, 94
Langland, E. and Gove, W. 34
Law Society 4, 37, 137, 193
Law Society's Gazette 131, 136
Le Corbusier 51
Leoff, C. 11
Levin, I. 129
Leevers, K. 146

Levison, D. and Atkins, J. 38, 132, 144, 146
Lewis, R. and Maude, A. 37
Leybourn, W. 43
Little, J. 3, 27
Loban, G. 33
Lorber, J. 8, 37, 138
LPAS 172
Lury, C. 14

Maccoby, E. 111, 121
MacDonald, M. 32, 46
McDowell, L. 23, 27, 52; McDowell, L. and Peake, L. 97
MacFarlane, A. 43
McLoughlin, J. 27
Macey, J. 66
Mahoney, P. 30, 33
Marcus, S. 37, 66
Marks, P. 131
Markusen, A. 167
Marriot, O. 8, 68
Marshall, J. 4, 38
Marwick, A. 84
Marx, K. 24, 25, 56, 175
Massey, D. 5, 27
Matrix 11, 27
Mead, M. 24
Menzies, C. 9
Merrett, S. 44
Merton, R. 26, 49
Metcalf, A. and Humphries, M. 24
Midland Bank Trust Company v. Green 28
Miller, D. and Swanson, G. 25–6, 32, 37
Millerson, G. 37
Millett, K. 23
Mills, P. and Oliver, J. 45, 47
Mills, C. Wright 13, 31
Ministry of Education 72
Mitchell, J. 23
Molyneux, P. 131
Morgan, P. and Nott, S. 67
Morgan, D. 14
Morphet, J. 3
Morris, E. 67
Morris, T. 69

Name index

Mulford, W. 14
Mumford, L. 170

NAB 85
Nadin, V. and Jones, S. 37, 203
Napier, M. 45
Nationwide 178
NCVQ 85, 125
Norton Taylor, R. 44
Nott, S. 193

Oakley, A. 38
Okely, J. 15, 30, 32, 100
OPCS 3
Options 171
Orchard-Lisle, P. 66, 75

Pahl, R. 4, 5, 7, 69, 136, 149; Pahl, R. and Pahl, J. 33
Parkin, F. 6, 24
Pearce, L. 54
People 155
Percy Report 72
Perry, E. 62
Pevsner, N. 64
Pickup, L. 167
Pickvance, C. 4, 25
Pilkington Report 72
Pinch, S. 4
Potter, S. 167
Power, A. 61–3
Purvis, J. 15

Quoin 78

Rapoport, R. and Rapoport, R, 33
Ravetz, A. 64
RIBA (Royal Institute of British Architects) 8, 17, 203
Richards, A. 53
Richardson, B. 53
RICS (Royal Institution of Chartered Surveyors) 3, 7–9, 13, 17, 31, 39, 43–4, 48, 56, 60, 63, 64, 66, 69–70, 72–3, 78–9, 82, 85, 92–5, 100, 125–6, 128, 130, 132, 134, 157, 163, 174, 180, 190; Royal status 69

Riley, M. and Bailey, C. 143
Robbins Report 34, 72
Roberts, P. 67
Robinson, C. 4, 137
Robinson, E. 34, 72
Rowbotham, S. 10, 23
Rowe, M. 60
RTPI (Royal Town Planning Institute) 8, 17, 37, 60, 146, 153, 174, 192, 196, 203

Saffioti, H. 23
Saks, M. 31
Sarup, M. 30
Saunders, P. and Williams P. 163
Saunders, A. and Wilson, P. 37
Savage, W. 38
Sayer, A. 28
SBP 11, 110
Schuster Report 70
Sharpe, S. 24, 30
She 100
Sheffield 106
Shepherd, W. 66
Siltanen, J. and Stanworth, M. 4
Silverstone, R. and Ward, A. 32, 37
Simmie, J. 8, 24, 27, 69
SITE Report 71, 81, 85
Skeffington Report 69
Smart, C. 15, 37
Smith, M. 63, 78
Smyth, P. 82
Snow, J. 79
Sousby, J. 79
Spencer, A. and Podmore, D. 4, 37
Spender, D. 24, 115
Stacey, M. and Price, M. 23
'Stackup' 143
Stanley, L. 168; Stanley, L. and Wise, S. 13
Stanworth, M. 29, 34
Stapleton, T. and Netting, R. 33
Steel, R. 59
Stewart, K. 26, 51, 164
Stimpson, C. 3, 27, 167
Stone, J. 145
Strauss, A. 8
Strutt and Parker 46

Name index

Sturge, J. 45, 48–9, 57
Swenarton, M. 64
Swords-Isherwood, N. 4

Taylor, B. 67
Teale, M. 72
Thomas, K. 37
Thompson, FML. 7, 8, 17, 29, 43–50, 55–9, 61–2, 66, 69–71, 81
Thompson, J. 29
Times, The 171
Torre, S. 27
TPI 60
Trepas, B. 66
Turner, J. 84

Valin Pollen 100
Vallance, E. 77
Venables, P. 60

Wagner, P. 52
Walby, S. 23
Walford, G. 30
Walker, L. 50
Walsh, L. and Gibson, M. 37
Ward, C. 70
Ward, D. 76
Ware, V. 162
Wareing, N. 34, 74

Watson, J. 46; Watson Report 70
Weber, M. 6, 25, 27, 50
Weiner, G. 34
Weisman, L. and Birkby, N. 34
Wekerle, G. 27, 67, 139
Wells Report 70
Westergaard, J. and Resler, H. 31
WGPW 62
WGSG 3, 27, 67, 173
Whelan, C. 84
Whitburn, J. 34, 74
White, J. 38
Whitelegg, E. 51
Whittington, E. 89, 94, 102
Whyld, J. 34
Whyte, W. 32
Whyte, J. 4, 36, 38, 145, 192
Wigfall, P. 36–7
Williams, M. 38
Willis, A. 69
Willis, P. 29, 32, 39
Wilson, E. 67
Woodhall, M. 31
Woods, P. 14
Woodward, D. 38
Working Woman 100
Wright, O.E. 31, 33

Young, M. 56

231

Subject index

'A' levels 94
abstract woman 101, 171
abstraction 119
academia 34, 98
accountants/cy 4, 44, 71, 99, 124, 136, 189, 196, 202
administration, of courses 98
admissions 99
age 134, 153
agents/actors, in sociological theory 28, 186
agricultural colleges 58; land 67
airing cupboards 120
alliances 164
American business women 23
Alonzo Dawes 47
analytical thinking 103
ancient partner 144
anorak, orange 8, 154, 156
anthropological research 23
anti-academic, pro-practical 32
anticipatory socialisation 35
antiques 43
apprenticeship 56
Arab women 93
Archers 84
architects, 7, 56, 66, 72; women 97
architectural courses 102
architecture 50, 94, 163
art 29, 43, 102
Art Deco Building 163
arty-farty 163
ASI (Architects and Surveyors Institute) 19, 44, 203
Asian men 153

assertiveness 39, 196
associate partnerships 125, 135
Association of Women House Property Managers 63
atmosphere 106
attractive women 112, 134, 153
auctions 47, 139
Aunt Sally 184, 192
axe, chopping 155

babies 107, 145, 146; baby changing areas 172
badge, Lionesses 82
backcloth, male 12
banks 125
Barclay, Irene 61
'beer swilling yobos' 93
Big Bang 124
Bill of Quantities 69
biology 102, 145
Birkbeck College 59
birth control 145
black: people 116, 156, 185; female student 115; men 150; overseas students 92, 115; technicians 150; women surveyors 140, 204; black/'coloured' men 94; *see also* 93, 153
blancmange 182
blue print, of society 28
BMA (British Medical Association) 202
boat owning fraternity 170
bookish student 103
bourgeois feminist 9, 22, 61, 79,

Subject index

154; sociology 25
bourgeoisie 52; female 52; *petite* 25
breadwinners 191
bricklaying 30
bricks 141
'bricks and mortar' 162
Bristol 174
Broadwater Farm estate 162
brochures, promotional 173
brothers 110–11, 188
brutalistic architecture 106
builders 65
building: business 46; correspondent, woman 83; site 14, 100, 150; societies 125; surveyors 93, 95, 123; women 136
built environment 3, 10, 12, 64, 181
Bulgaria 7
bureaucratic cultures 26; personality 26, 49
business studies 36, 102, 106, 119; suits 153, *see* clothes; women 4
businesslike 181

Cadbury 60
cadets, district valuation 137
cakes, and concrete 120
calves, at market 150
Cambridge 46, 58
canoe, paddle own 23, 107, 178
capitalism 28, 43
capitalist 49, 164; feminist 22
caprice 91, 149, 183
cars 68, 118, 151, 156, 167–8, 171; allowance 135, 156; telephone 129
career advisors 100; breaks 146; strategies 183; talks 101; woman 38
CareerTrack 38
caring, for elderly 193
caste 93
Castle, Barbara 76
castle, English man's home 53
cattle 139
Cenotaph, War Memorial 69
centenary, RICS 78
central business district, CBD 166
central government property

agencies 128
certificate for women housing managers 62
Chadwick 55
chain stores 128
chain surveying 47
chairperson 115
chameleon like profession 72
channelling of women 83, 105
'chaps' 104
Chartered Surveyors Institution 69
Chelsea pensioners 77
Chicago school of sociology 13
child: childcare 32, 81, 117, 191, 193; children 67, 82, 133, 137, 145, 171; minder 26, 108; molestation 117; rearing 84; too many 116; training 25
Chinese stamp 78
cigarette packets 49
cigarettes 106, 113
CIOB (Chartered Institute of Building) 203
civil engineering 50
civilisation 43
classics 59
classroom interaction 33, 113
clerical staff 130
client 58, 66, 148, 170, 176; corporate 125
closure 6, 24, 105, 183
clothes/dress 106, 153, *see* suits
Clutton 56; John 57; William 45
CNAA (Council for National Academic Awards) 90
coin, two sides of 26, 188
co-operative housekeeping 54
coal merchant 47
cognate degree 73
cognate non-chaps 104
College of Estate Management (the CEM) 60, 74, 207
collusion 147
colonial approach 162
colonialism 116
colonisation 48, 147
'comfortable' 6
commercial market 68; property 125,

233

Subject index

141; world 174
commercialism 44
committee members, women 82
common sense view 48
community 161–2; facilities 119, 169; groups 68
company cars 167
computers 36, 102, 119, 124, 134, 144, 157, 187; grid format 69; home, computers 157
concrete 114, 120
conditions of planning permission 172
conferences 197
conferred, professional socialisation 35
confirmed, professional socialisation 35
conflict, women in education 36, 182
conscience 186
conservation 163
conservative 49, 66, 70
construction 97, 106, 119, 120
construction industry subculture 128
consumerism 52
consumption 65, 76
contractors 130
control 27
cookery 29, 100
Corbusier, Le 51, 64
corporate structure 125
correspondence courses 58, 71, 73, 95, 97
corridors 106, 121
cost factor 89
council estates 115; flats 118; tenants 77
councillors 7
couples, out of order 145
CPD (continuing professional development) 71, 84, 196
creches 11, 108, 169
credentialisation 30, 71, 91
cricket 80, 151
crisis, in profession 85
crit session, assessment 112
cultural capital 105; lag 186, 188
culture shock 110; spongy 35

dangerous, urban environment 84
daughters of surveyors 46, 149
Dawes, Margery 46
decentralisation 174
decry 'theory' 33
Deeds of Partnership 137
deference 28
deferred gratification 78
demographic time bomb 190
design, women's needs 180
destination, of graduates 104
development 119; portfolios 163; process 7, 174, 181; women and 176
Diary, Jennifer's 80
different attributes 197
different policies? 176
diploma disease 91
'dirty words' 35
dispersed ethnography 15
District Surveyor 55
District Valuation Office 131, 138
district valuer, woman 82
divorce 28, 146
dog pounds 138
dogsbody 147
domestic science 102
double-consciousness 10, 13
double standards 103
draft dodging 68
dragon 77
drama, school subject 102
draughtsmanship 97
drawing office 70
dress 106, 110, 133, 153, 168, *see* clothes, suits
Duke of Edinburgh Gold Medal 102
dynasties, surveying 35, 46, 141

economics 35, 36, 119
Edinburgh 127
education, sociology of 29
Education Acts: 1870 57; 1902 57; 1944 70
educational policy 70
efficiency, and modern architecture 51
Eighties, the 80

Subject index

elderly, the 161
elderly men in the profession 50
elites 31, 46
Elton, Lord 142
embroidery, school subject 102
emotional 6
emotional housework 73
Empire, British 52
employment 173
Enclosure Acts 47
engineering courses 94
'English' 113
enterprise culture 10, 36, 80, 189
entrepreneurial cultures 26
entry requirements 51
equal opportunities 90, 130, 137
essentialist (all lads are bad) 17
estate agency (realtors) 9, 83, 93, 95, 125, 126, 128, 139, 153, 155; management 60, 94, 102, 119; management students 3; managers 50, 174
ethnic minority women 150, see black people
ethnographic methods 13–15
eugenics movement 54
Europe 8, 51, 196
evangelical reformism 54
examinations 70
exclusionary mechanism 47, 71
exploitation, class 31
extended surveying families 141

fabianism 60
factories 51, 67, 173
faculty 106
falling rolls 182
False Dawn 86
family 25, 112, 168
Far Eastern women 93
farming backgrounds 46
father to son 47
father's firm 134
fax machines 195
fee earning capacity 137
fee sharing 135
'feminism' 20, 84, 176, 179; bourgeois 22; capitalist 22; first wave 52; material 54; reformist 22; second wave 185; socialist 23; urban 52, 172
feminist planners 172; fallout 172; orthodoxy 189
femininity 153
femocrats 11
feudalism 44, 72
FIG (International Federation of Land Surveyors) 7, 78
'figures in the landscape' 27
film, RICS Centenary 78
filtering 182
'finders, minders and grinders' 133
Finland 7
First Aid 113
first causes, sociological 11, 184
First Wave of Feminism 52
first woman 81; JO chair 80
flat, shared (apartment) 132
flexi-time 194
flexibility 191
football 118
foreign clients 154
foundations, site 141
fractions, professional classes 24
fragmentation 126, 165, 183
fraternity 129, 161, 174–5
freedom, in profession 35, 194
Freemasons Tavern 56
Friends meeting house 49
'fun', shopping (marketing) as 170
functionalism, architecture 64

garages 168
garden, house (yard) 163
garden cities 60
gas and water socialism 54
gatekeeping 24
gauche, attitudes 29
gays 156
general practice 73, 94, 123–4
gentlemanly 106
gentlemen's club image 72
gentrification 167
geodesy 92
geographer 27, 163; woman 11, 164
geographical determinism 165

Subject index

geography 97, 102, 119
'get rich quick students' 133
'getting out and about' 139, 194
'girl friendliness' 192
girlie calendars 144
Glasgow 127
glue, course cohesion 85
golf 108; courses 169
'good words' and 'dirty words' 35
grades, student 101–2
graduate school 30
grandfathers 57, 141
Great Fire of London 45
GLC (Greater London Council) 80, 172
green belts 166
green field sites 52, 174
guilds 47

hair 97
handwriting 112
hard-hat roles 81, 100
harmonisation (EC) 196
Harvard Business School 12
Head of Department level 97
headaches 146
headed note-paper 145
heating and ventilating engineers 101
heavy equipment 100
height, applicants 131
helper 115, 134; helpmeet 46, 111, 114, 139–40, 147, 176
Heroes 64
heiresses 52
high heels 153
high street banks 193
Hill, Octavia 55, 63
Hoddell Pritchard, 46
home commitments 126
Home Counties 126–7
'home of your own' 163
home/work linkages 32
horizontal distribution 138
horse riding 139
hostelry 143, *see* pubs
housewife 116, 118, 171, 186; militant 121
housing 80, 110, 132, 163; Housing Act, 1919 64; courses 94; crisis 179; management 62, 162; manager 9, 55, 78; Mangers Certificate 63; Working Party 38; woman 76, 82
Howard, Ebenezer 54
HQ, RICS 44, 77, 79
humanities 108
hunter, food gatherer and 170
husband and wife team 84
husbands of surveyors 126
hydrographic surveyor 81
hypermarket 174

immature, perceived as 80
impersonal ethos 107, 116
incidental men 129
indemnity insurance 136
indispensible women 132
individual energy 56
individualism, possessive 43
industrial property 140
Industrial Revolution 50, 55, 57
infrastructural services 51
Inner Areas Act, 1979 167
Inner City 69, 129
innuendo 83–4
insignificant, numbers 90
Institute of Chartered Accountants 202
Institute of Civil Engineers 204
Institute of Housing 63, 146, 203
Institute of Housing Managers 64
insurance and pension funds 130
insurance companies 127
inter-personal relationships 33
interviews 133
investment accountants 124; brokers 8; portfolios 43
invisible, women 62, 115, 173
Ireland 127
ISVA (Incorporated Society of Valuers and Auctioneers) 203

jeans 110
Jewish business women 152; families 49
JO (Junior Organisation) 123, *see*

Name index
job-sharing 144, 146
Joseph's findings 34–7
journalists, women 82–3
jumpers and skirts 154

King's surveyors 47
kiss 144

Labour Government 49, 65, 70; labour movement 173
ladies literary society 57
ladies, presence of 69
lady bountifuls 9, 53
lady gaffer 151
Lamplugh Suzy 85, 156
land 8, 17, 35, 52, 180; agency 81, 123; management 43, 188; -owners 125; survey 119; surveying 3, 47, 92, 102, 187; surveyor 50, 68; Surveyors Club 56; surveyor's wife 49
land use and development 180
landed interests 175; professions 37; property 26
landlord and tenant 63, 141
landscaping 140
Landseer 84
late developers 37
lavatories (toilets) 77, 79, 106, 171, *also see* public conveniences
law 4, 8, 37, 52, 119, 121, 189
Law Society 202
lawyers 29, 56, 71, 99, 124, 196; monstrous regiment of 44; women 97
Lays of Ancient Rome 74
Le Corbusier 51
leapfrogging 135
lecturers 157; as failures 98
left wing 163
legal contract 140, 176
leisure 169
levels, sociological 20, 180
liberal feminism 22
librarians 98
lighting 176
Lincolnshire 127

linguistic ability 154
Lionesses 82, 84, 128
Lions 136
'listen', to the people 161
livestock 43, 139
local authority 96, 128
local government fraternity 144
London 25, 36, 47, 91, 95, 127, 131, 140; Docklands 167; Fire 45; practices 125; University 60
longitudinal studies 38
'lowering the standard' 108
lunch 98

magazines 100
'make the familiar strange' 28
male backcloth 12; bonding 26; gatekeepers 184; preserves 187; proletariat 52; working class 29
'malestream' 4
management 8, 37, 119, 144; structures 37
Manchester 127
manpower 4, 70, 125, 133, 155, 184; theory 31
map reader 111
map-making 187
marginal: men 46; women 152
marinas 170
market 119, 181; demand 175
marriage broker 45; markets 45
married women's property rights 52
marxian literature 24–5
masons 47
mass production economy 64
maternal deprivation 26
maternity leave 23
mathematical models 79
mathematics 102, 120
mature students 85
mechanisms, within profession 4, 105
medical profession 4, 8, 37, 85, 202
meeting, formality of 107
meritocracy 56, 99
Merry Hill shopping centre near Dudley 170
Merseyside 127
Metropolitan Building Act, 1844 55

Subject index

middle-class women 74
Midlands 91, 127
'milk round' recruitment 131
minerals surveying 93, 123, 140
mini skirts 78, 153
'Miss' 90
Miss World 83
mixed schools 33, 91
modern valuation techniques 36
money, women's attitude to 177–8
monkeys, zoo 143
morgageriat 44
mortgage market 178
mothers 99
mothers and sisters 186
motorway 8; construction 141; intersections 174
murder 176

NAB (National Advisory Body) 85
nannies 26, 146
National Association of Estate Agents 203
naughty, attitude 36, 100
negotiating 151
negotiators 95
neighbourhood unit 67
nested hierarchies 25
networking 197
neutrality, professional 3
New Towns 67
'New Woman' 62
nicenesses and nastinesses 6, 38
Nigeria 7
non-cognate graduates 95, 104
'normal' student 34
North America 8, 48, 72
Northern England 127
novelty, women as 80
nuclear power station 138
nurseries 66

'O' level land surveying 102
occupational psychology 38
offices 143, 164, 173; office blocks 43; office ladies 136; office staff 79, 98; office women 25, 187, see typists

officer core 78
older working women 190
One England 189
'ordinary' parents 99
Ordnance Survey 48, 68
out and about 28, 36, 155
out-of-town shopping centres 164
overseas student 112
Oxbridge 58, 96

panel beating 114
parapets 110
parent 25, see mothers
Parliament 135
part time 144, 196; female labour 67; male 'part-timer' 98
partners 74, 135
partnership 104, 154
paternalistic, attitudes 185
patriarchal women 90
patriarchy 11, 28, 187
pecking order 59, 113
pension funds 130
'peoplelessness' 119
Perry, Evelyn 61
person selection 26
personal factors 187; views 37
personnel officer 129
perspective, professional 35
petite bourgeoisie 25
philanthropic factory owners 51
photograph, misleading 14, 101
photogrammetry 92
planimeter 148
planners, women 11, 172
planning 50; and development 123–4, 164; law 172; 'planning is for people' 69
Player, John 45, 49
Player Sturge, Jacob 45
polytechnics 34, 72–4, 91, 192; first wave 60
ponies 139
poor, look 157
porters 45, 151
post-graduate 94–6
Post War Reconstruction 65
power dressing 39

Subject index

practicality 35
precision 35, 102, 107
pregnant 145
President of the RICS 46, 79
prison construction 138
private consultancy 98; practice 130; property law 119
professional: careers 37; competance 51; demeanour, impersonal 177; expertise 187; office 71; persona 116; socialisation 7, 24, 35
professionalism 35, 107
professions 4, 37; sociology of 31
project work 110–11
property: wives as 27; boom 69, 71; development process 175; fraternity 174; management 139, 176; professional 124; research 139; women and 177, 197
prospectus 100
prostitution 117, 156
provinces 136
provincial firms 125
public conveniences (toilets) 77, 171, 172, 176, *see also* lavatories
public school: girls 30; boys 59, 95; thickos 99; white males 115
public sector 130–1, 138
public transport 77, 80, 168
pubs 117; culture 109
pussy cat bows 153
pyramidic structure 185

Quakers 48–9
quantity surveying 50, 56, 83, 93, 106, 123–4
quasi–professions 32
Queen 76
questionnaires 35
'quite by chance' 141
quota system 93

racing cars 49
railways 50, 53
rating and valuation work 141; Rating and Valuations Association 203
Reading 60, 207

'real jobs' 173
realism, theory 28
realtors 8, *see* estate agents
reasonable return 170
recruitment brochures 131; crisis 136
Red Nose Day 163
referee 51
refrigerator 76
regional location 106; planning 119
relatives 146
rent reviews 141
research 124
researcher 14
residential development 126, 138, 164
retail centres 43; development 170
retirement 193
retrospective ethnography 15
right type 9, 25, 183
right wing 162–3
Rochdale 168
role 182; models 84, 186
roughness 30
rudeness 112
rugby 39, 108–9, 175
'rule of thumb' 57
rural: areas 143; estate 45; estates 27; practice 139

'safe as houses' 54
safety 80
safety net 94
sailing 169
salaries 69, 132
'salvation by bricks' 53
'same' 103
sandwich course 74
satellite geodesy 188
Savill, Maria 46, 50
school 29–30, 146; holidays 194; leavers 4, 189; teachers, women 100
science 4, 37, 50, 100, 102, 144
Science Parks 141, 164
Scotland 127
script, for professional women 4
seagulls 150
Sea Lady 81

239

Subject index

seats, in shopping centres 171
Second Wave of Feminism 185
secretary 129, 149; male 147; role 111
security guards 148
self defence 154–5
selfish 53–4, 109, 177
senior partners 125
seniority 104, 134
sensitising concepts 29
'seriousness' 101
service delivery 162
settled land 43, 121
Seventies, the 79
sewers 61, 179; and drains feminism 54
Sex Disqualification (Removal) Act, 1919 17, 61
sexism 78
sexist comments 164; stereotypes 121
sexuality 24
Sharp, Evelyn 76
shooting rights 43
shopkeepers 32, 170
shopper 168
shopping (marketing) 65–6, 76, 79, 170; local 118
short hair 97
Shula 84
SIFT (Surveying Information for Tenants) 162
silence 38
silly 73, 114, 161
sin 49
'sister's an architect' 121
sites 48
Sixties, the 77
skin care 77
sloanes 184
slow-witted men 55
slum clearance 66
slum dwellers 62
slurry pit 45
Smith, Mary 63, 78
Smith, Miss M.V. 59
smoke, cigarette 113
social aspects 113, 119; issues 174; trends 161

'socialist' 162, 164; government 66
socialist-feminist perspective 23
Society of Women Housing Estate Managers 63
sociology of education 5, 13
socks 111
sole practices 125
solicitors 56, 137; women 130
South England 127
South East England 131
South West England 127
spatial and aspatial 5
spatial fetishism 52
'special' 180; needs 172
specialist 184
spectrum 8
splutter effect 91, 137, 182
spoil heap reclamation 140
sport 29, 102, 108–9, 112, 117, 131, 163, 169, 174
spring line settlement 165
squash 109; courts 169
staff meetings 114
standards 107
Stapleton, Beatrice 59
state intervention 53, 64
state schools 29
Stepford Wives 129
stereotypes 186
straight 10
stream of consciousness 16
structure and agent 28, 186
Strutt, Edward 46
Students Union 106
Sturge 48
sub-proletariat 32
subuniverse 11, 56, 181
subculture 5–6
subjects taught 34
suburban hareem 52
suburbanisation 67
suburbia 65
success 35
Suffolk 127
suit 132; business 171, *see* dress, clothes
sullenness 36
superannuation 137

240

Subject index

supermarket car park 168
surplus capital 52
surveying: dynasties 185; education 3, 33, 69; education, standards 85; pole 48; science 92; spectrum 8
surveyor's daughters 78; girl friends 82; wives 82
Surveyors Institution 59, 69
swearing 144
sweetheart 46
symbolic interactionism 25
systems theory 27

talent 135
talisman, working class as 179
tapestry of life 185
taxation 49
technical colleges 60; northern 95
technical drawing 102
technician grades 124
technicians 98, 148
technology 4, 50, 141, 180; more male than science 37
telephone conversations 91, 129
test tubes 117
textbooks 55
thatcher clones 154, 184
theodolite 8, 50, 77, 100
Tithe Commutation Act 48
toilets *see* lavatories and public conveniences
Town and Country Planning Act, 1971 69
town centre redevelopment 66, 68
town planner 7, 13, 24, 37, 48, 50–1, 65–6, 71, 99, 196
town planning 60, 94, 97, 110, 112, 117–20, 162–3, 174; Chair of 60; Institute (TPI, now the RTPI) 60, *see* Name index
TPC (test of professional competence) 90
tracers 70
transmission 30
transmitter 33
transport 76
Transport Board 128
transportation 165, 167

triangulating 15, 68
trolley, supermarket 171
trouble 185
turnstiles 77
tutorial 114, 117
typist 98, 147, 157
tyranny of the office 195

uncomplicated people 181
underground car park 176
unemployed 116
upper-class women 53
urban economics 119; feminist literature 80; feminist movement 68, 77; managers 69; sociology 4; sprawl 67
Use Classes 172; BI use 172
utopianism 54

valuations 3, 97, 119–20, 144, 176
vertical distribution 5, 134
vices 58
visual spatial abilities 111, 120

Wales 127
wally 120
war service 70
Washington, George 48
'watching cricket' 80
waterproof overalls 154
'we' 113
wealth 24
weather 109
weight 131
wellies 125, 154
Wells, William 60
West End practices 127
white heat of technology 72
'who gets what, where, and why?' 4
wines 47
WIP (Women in Property) 197
wives 27, 129, 149
'wogs and women' 94
women: committees 80; engineer syndrome 81, 144; housing managers 78; property developers 129; solicitors 193; 'women only' courses 101; womentalk 25

Subject index

wordprocessors 157
work twice as hard 34
working class 27, 58, 166; boys 74; housing estates 61; women 53
women and men 167
written reports 112

young boys 50
yuppie 85
yuppification 124, 167

zoning 51, 166